15 | Springer Series in Chemical Physics
Edited by R. Gomer

Springer Series in Chemical Physics

Editors: V. I. Goldanskii   R. Gomer   F. P. Schäfer   J. P. Toennies

Vibrational Spectroscopy of Adsorbates

Editor: R. F. Willis

With Contributions by
B. K. Agrawal   G. Allan   J. A. Creighton
B. Djafari-Rouhani   L. Dobrzynski   B. Feuerbacher
P. Hollins   D. A. King   D. M. Newns
J. Pritchard   N. Sheppard   D. G. Walmsley
R. F. Willis   C. J. Wright

With 97 Figures

Springer-Verlag Berlin Heidelberg New York 1980

Dr. Roy F. Willis

Cavendish Laboratory, University of Cambridge,
Cambridge CB3 OHE, England

Series Editors

Professor Vitalii I. Goldanskii

Institute of Chemical Physics
Academy of Sciences
Vorobyevskoye Chaussee 2-b
Moscow V-334, USSR

Professor Robert Gomer

The James Franck Institute
The University of Chicago
5640 Ellis Avenue
Chicago, IL 60637, USA

Professor Dr. Fritz Peter Schäfer

Max-Planck-Institut für
Biophysikalische Chemie
D-3400 Göttingen-Nikolausberg
Fed. Rep. of Germany

Professor Dr. J. Peter Toennies

Max-Planck-Institut für Strömungsforschung
Böttingerstraße 6–8
D-3400 Göttingen
Fed. Rep. of Germany

ISBN-13: 978-3-642-88646-1   e-ISBN-13: 978-3-642-88644-7
DOI: 10.1007/978-3-642-88644-7

Offset printing: Beltz Offsetdruck, Hemsbach/Bergstr. Bookbinding: J. Schäffer oHG, Grünstadt.
2153/3130-543210

Preface

Over the past few years, there has been a growing awareness of the vibrational properties of solid surfaces and adsorbates due to a steady growth in
the number of experimental techniques which have evolved with sufficient
resolution and surface sensitivity. An understanding of the surface vibrational modes is of fundamental importance in many areas of the physics and
chemistry of surfaces, most notably in the field of heterogeneous catalysis
on metals and alloys.

The present volume derives from a one day meeting of invited lectures,
held under the auspices of the Thin Films and Surfaces Section of the Institute of Physics in the Cavendish Laboratory, University of Cambridge,
13 December 1979. The object was to bring together specialists from various
diverse fields who would examine the wide variety of methods currently available for studying surface adsorbate vibrations. Since these methods cover
several scientific disciplines, it was subsequently felt that it would be
useful to provide a permanent record of the talks as a source for future
reference by workers in what is rapidly becoming an expanding field of interest in an increasing number of laboratories. The contributions, however, are
not in any way meant to constitute exhaustive reviews. Rather the effort has
been directed more towards providing an introduction to the already extensive
literature which exists in each subject field, stressing more the fundamental
principles underscoring the various techniques, current problem areas, future
directions, etc., so as to reflect the seminar nature of the original proceedings.

Following a brief introduction (Chapter 1), electron-energy-loss spectroscopy, which has provided enormous impetus to this field, is introduced in
Chapters 2 and 3. The importance of dipole selection rules for scattering in
the specularly reflected electron beam direction is emphasized in Chapter 2,
with angle and impact-energy dependent studies discussed in Chapter 3. The

collective nature of adsorbate-induced surface (optical) phonon modes is described in Chapter 4. The methods of inelastic electron tunnelling spectroscopy of adsorbate vibrations (Chapter 5), inelastic molecular beam scattering (Chapter 6), and inelastic neutron scattering (Chapter 7) studies of surfaces provide additional complimentary information. The optical techniques of infrared reflectance absorption spectroscopy and surface enhanced Raman spectroscopy are described in Chapters 8 and 9, respectively. The relationship between the vibrational modes of adsorbate species in metal cluster compounds and complexes is compared with similar vibrations of adsorbates on metal surfaces (Chapter 10).Finally, a specific case illustrating the importance of vibrational coupling between adsorbates viz. carbon monoxide adsorbed on Pt(100) and Pt(111) surfaces is described (Chapter 11).

The number of adsorbate systems described has been purposely limited, examples being drawn from an extensive literature only in so far as to illustrate the principles behind each method. In this way, it is hoped that this collection of articles will serve its main purpose of introducing readers to current methods of research into the vibrational properties of adsorbates.

Cambridge, July 1980   *Roy F. Willis*

Contents

List of Contributors

Agrawal, Bal K.
    Department of Physics, Allahabad University, Allahabad, India

Allan, Guey
    Laboratoire d'Etude des Surfaces et Interfaces (associé au CNRS) I.S.E.N.
    3, rue F. Baës, F-59046 Lille Cêdex, and
    Laboratoire de Physico-Chimie, Université Claude Bernard,
    43, Bd. du 11 Novembre 1918, F-69622 Villeurbanne Cêdex, France

Creighton, J. Alan
    Chemical Laboratories, University of Kent, Canterbury CT2 7NH, England

Djafari-Rouhani, Bahran
    Laboratoire d'Etude des Surfaces et Interfaces (associé au CNRS) I.S.E.N.
    3, rue F. Baës, F-59046 Lille Cêdex, and
    Laboratoire de Physico-Chimie, Université Claude Bernard,
    43, Bd. du 11 Novembre 1918, F-69622 Villeurbanne Cêdex, France

Dobrzynski, Leonard
    Laboratoire d'Etude des Surfaces et Interfaces (associé au CNRS) I.S.E.N.
    3, rue F. Baës, F-59046 Lille Cêdex, and
    Laboratoire de Physico-Chimie, Université Claude Bernard,
    43, Bd. du 11 Novembre 1918, F-69622 Villeurbanne Cêdex, France

Feuerbacher, Berndt
    Astronomy Division, Space Science Department of ESA ESTEC
    Noordwijk, The Netherlands

Hollins, Peter
    Chemistry Department, Queen Mary College, London E1 4NS, England

King, David A.
    The Donnan Laboratories, The University, Liverpool L69 3BX, England

Newns, Denis M.
    Institute of Theoretical Physics, Chalmers University of Technology,
    S-41296 Göteborg, Sweden
    Permanent address: Department of Mathematics, Imperial College,
    London SW 7, England

Pritchard, John

    Chemistry Department, Queen Mary College, London E1 4NS, England

Sheppard, Norman

    School of Chemical Sciences, University of East Anglia,
Norwich, NR4 7TJ, England

Walmsley, D. George

    School of Physical Sciences, New University of Ulster,
Coleraine, Northern Ireland

Willis, Roy F.

    Cavendish Laboratory, University of Cambridge, Cambridge CB3 OHE, England

Wright, Christopher J.

    A.E.R.E. Harwell, Didcot, Oxon, England

1. Introduction

Roy F. Willis

Infrared absorption vibrational spectroscopy has been used for many years
to identify bonding arrangements in molecules [1]. Each bond has its own
characteristic frequency so that the vibrational spectrum tells us much about
molecular structure. Similar information can be obtained when molecules
are adsorbed on solid surfaces [2]. For example, if carbon monoxide is adsorbed, sometimes one observes the individual vibration modes of dissociated
C and O. Other times one observes the CO stretching mode indicating that
the molecule remains undissociated. Each of these vibration modes (C, O
and CO) has a different frequency for each bonding site. Also, the spectral
intensities relate to the concentration of the various adsorbed species.
Thus, vibrational spectroscopy of adsorbates is a powerful means for identifying the type and concentration of molecules adsorbed on solid surfaces.
This is of great importance in the field of heterogeneous catalysis.

1.1 Vibrational Spectroscopy in Relation to Surface Science

There exists an enormous literature describing the vibrational properties
of adsorbed molecules [2]. However, until a few years ago this was almost
exclusively limited to infrared transmission spectroscopy of finely divided
substances. This made it difficult to relate the observed vibrational frequencies to adsorption on any specific site due to the ill-defined nature of
the substrate surfaces. The trend over the past few years, therefore, has
been to develop methods of vibrational spectroscopy, compatible with other
surface analytical techniques, which can be applied to well-characterized
adsorbate monolayers on single crystal (usually metal) surfaces. For
example, when a simple diatomic molecule such as CO is adsorbed, the vibrational frequency of its stretching mode indicates whether the molecule
sits on one or two (rarely more) surface sites. On a simple cubic crystal
surface with four-fold symmetry (such as the (100) low index plane), there
are four possible high symmetry sites, Fig.1. Low energy electron diffraction (LEED) helps us to identify the overall crystallographic arrangement,
and often this is sufficient to identify the actual bonding site(s) on the
surface [3]. This is a first fundamental requirement if we are to understand
the detailed nature of the bonding between the molecule and the surface.

Unfortunately, the identification of the bonding site is not always
unambiguous from the LEED pattern. The task is complicated if, for example,
the adsorbed layer is out-of-registry with the substrate lattice, or worse,
if it is disordered. However, if it were possible to observe *all* of the
vibration modes, then the *number* of non-degenerate vibrations is sufficient
to establish unequivocally the point group symmetry of the adsorption site
[4]. The actual number of discrete vibrations likely to be observed in

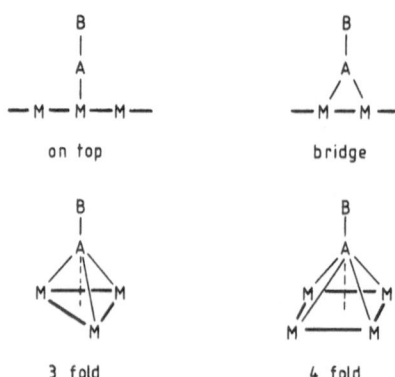

Fig.1 Showing four possible site symmetries for the bonding of a diatomic molecule AB to metal surface atoms M.

any spectrum is, however, dependent on certain "selection rules" inherent in the excitation mechanism relating to the particular vibrational spectroscopic technique employed. These selection rules determine the spectral intensities of the vibration modes, an understanding of which is fundamental for the problem of surface site geometry and structure.

1.2 Techniques in Surface Vibrational Spectroscopy

There has been a growing awareness in recent years of the vibrational properties of surfaces and adsorbed layers due to a steady growth in the number of experimental techniques which have been developed with sufficient resolution and surface sensitivity (re: Table 1). These techniques may be catagorized in terms of the nature of the exciting probe radiation - electron, photon and neutral particles. Foremost among these has been the development and widespread use of electron-energy-loss spectroscopy (EELS) which, together with infrared-reflection-absorption spectroscopy (IRAS) of single crystal metal surfaces, has provided enormous impetus to the field. Closely related in terms of the principles underlying these two methods, we have inelastic-electron-tunnelling spectroscopy (IETS) and surface enhanced Raman spectroscopy (SERS).

The selection rules operative in these techniques are similar in that the vibrational transition moment involves the interaction of the adsorbed molecules with electromagnetic fields induced at metal surfaces. In the case of IRAS, the optical theory is fully developed and predicts strong absorption only if there is a net dipole moment change normal to the metal surface [5]. This is a consequence of the fact that the reflectivity is high at infrared frequencies and the field outside the surface has no parallel component. A similar selection rule also operates in EELS with the restriction that the scattering be in the specularly reflected electron beam direction [6]. Off-specular, this "dipole" selection rule breaks down and vibration modes parallel to the surface are observed [7]. This is an important aspect of the electron scattering method: EELS offers the considerable advantage over IRAS of being able to distinguish between modes vibrating parallel from modes vibrating normal to the surface. Dipole scattering is a consequence of the *long range* part of the Coulombic interaction potential; off-specular scattering is due to the *short range* scattering potential of the ion cores of the substrate/adsorbate lattice.

Table 1 Techniques in Surface Vibrational Spectroscopy

Characteristic / Technique	IRAS	SERS	EELS	IETS	NIS	AIS
RESOLUTION (FWHM) cm^{-1}	1 – 5	1 – 5	25 – 100	1 – 5	50 – 75	1 – 5
SPECTRAL RANGE (approx) cm^{-1}	1500 – 4000	1 – 4000	240 – 5000+	240 – 8000	16 – 1600+	80 – 500
SENSITIVITY (% monolayer)	0.1	0.1	0.1	0.1	0.1	0.1
SAMPLE AREA (mm2)	10	1	1	1	v. large	1
SUBSTRATE	Crystals Metals and Insulators	Roughened Ag, Cu, Au	Crystals Conducting	Oxide-metal	Finely divided Metals and Insulators	Crystals Metals and Insulators
ADSORBATE	Few mainly CO	Few CN^-, Pyridine	Many	Many	Mainly H and other light atoms	None to date
AMBIENT PRESSURE	≤ 1 atm	Sample preparation ≤ 1 atm	< 10^{-6} Torr	Sample preparation ≤ 1 atm	≤ several atm	≤ 10^{-6} Torr
UHV COMPATIBLE	✓	✓	✓	✓	✓	✓
THEORETICAL SITUATION	Classical	Evolving	Evolving	Evolving	Developed	Evolving

The distinction becomes less clear in IETS where the electrons tunnel between two metal electrodes, the spacing of which is of a dimension comparable to that of the larger organic molecules which are usually adsorbed within the oxide interface [8] . Here the need for a fully quantitative theory of low-energy electron impact vibrational excitation of adsorbed molecules is particularly desirable at the present time. Infrared, as well as Raman, active modes are observed for reasons which have yet to be clarified. Similarly, the surface-enhanced-Raman effect is understood only in very qualitative terms [9]. Also, this recently discovered phenomenon is far more specific in that it relates to the observation of the so-called "giant" Raman resonances observed for molecules such as pyridine and cyanide adsorbed on specially prepared noble metal (principally silver) surfaces. The present consensus of opinion [10] is that the effect is a consequence of the complex nature of the electrodynamic coupling of the vibrationally excited molecules to the surface, and the (short range) interaction between the fields due to the polarized molecules and the locally induced surface electric field. Again, infrared and Raman active modes are observed.

The importance of the SERS method lies not so much in its possible future widespread application but in the fact that it has served to highlight the importance of microscopic electrodynamic processes associated with the vibrational coupling of molecules to metal surfaces. Such effects will play a determinant role in the reaction dynamics and kinetics of catalytic processes [11].

In contrast to the above methods, neutron inelastic scattering (NIS) [12] and atom inelastic scattering (AIS) [13] spectroscopies involve the interaction of neutral particles with surfaces, and do not require any knowledge of changes in the electric dipole moment due to changes in the nuclear positions of the vibrating atoms. Intensities are, in principle, exactly calculable, and it is possible to measure momentum transfers of the order of the magnitude of reciprocal lattice vectors appropriate to ordered surface layers. AIS in particular promises to become an important technique in this respect in view of the lack of any penetration of the low energy atomic beams (principally He) into the bulk lattice. With this method, it is possible to excite the collective vibration modes of the surface and to thereby determine surface phonon dispersion relations which relate to the vibrational coupling between the surface atoms [14].

1.3 Vibrational Coupling and Surface Phonon Modes

It is not possible in an experiment to observe the vibration modes of a single adsorbate molecule; the spectra are usually recorded as a function of surface coverage. With its superior resolution (compared with EELS), IRAS has revealed that the adsorbate frequencies shift considerably a) from their gas phase values, and b) with increasing surface coverage up to a saturation value. A coverage dependent property clearly points to the occurrence of coupled (or collective) vibrational behaviour within the adsorbed layer itself. The frequency shift as a function of wave vector provides information on the force constants between the adsorbed molecules [14]. The frequency shift from the gas phase value is indicative of chemical bonding changes due to chemisorption on the surface. The difficulty is in trying to separate these two effects (chemical bonding vs. vibrational coupling) in order to relate the observed vibrational frequencies of adsorbed species with similar vibrations in metal cluster compounds and complexes, whose bonding geometry is firmly established [15].

One approach which has been developed [16] is that of varying the iso-tropic ratio of ^{12}CO and ^{13}CO at different fixed coverages and observing the frequency shifts of ^{12}CO and ^{13}CO stretching vibrations. Assuming the molecules to interact as coupled dipoles, neighbouring molecules can vibrate either in-phase or out-of-phase with each other. If both molecules (i.e. di-poles) are of the same isotopic species, only the in-phase mode remains infrared active. However, if the molecules are of different isotopic composition, both the in-phase and out-of-phase vibrations become infrared active. According to model dipole-dipole coupling calculations [17], a system composed of a ^{12}CO molecule completely surrounded by ^{13}CO molecules absorbes at the frequency of the in-phase mode very closely (within 14 cm^{-1}) to the ^{12}CO singleton frequency. On the other hand, the frequency of the out-of-phase mode is close to the ^{13}CO singleton frequency. The difference in isotopic mass produces a frequency splitting of the order of 20 - 40 cm^{-1}. Frequency shifts of both bands are observed to occur as a function of isotopic composition and coverage.

Particularly striking examples of large frequency shifts occuring with increasing coverage are the systems CO/Pt(111) [16] and CO/Pd(100) [18]. In the latter case, it has been found that by varying the isotopic ratios of ^{12}CO:^{13}CO for different fixed coverages, it has been possible to determine the separate contributions to the overall coverage-dependent frequency obs-erved. That is, of the overall shift of 95 cm^{-1} observed for CO/Pt(100) up to a maximum coverage of 0.8 that of the surface density of substrate atoms, vibrational coupling accounts for 35 - 40 cm^{-1} shift, whereas changes in the chemical bonding energy are responsible for the remaining $\sim$ 55 cm^{-1} frequency variation. The effective force constant describing the interaction between two adjacent CO molecules would have to have a value of $\sim$ 0.10 m-dyne/Å to explain frequency shifts of this order of magnitude [19].

Such considerations are of importance in determining the lifetimes, frequencies and chemical bonding of adsorbates at metal surfaces. Vibrational energy can be dissipated via phonon processes within the surface and the substrate, or via dynamical charge transfer processes between the adsorbed molecules and the electronic states at the metal surface. We are just at the threshold of beginning to understand these processes, and the informa-tion provided by the methods described in this book holds promise that we shall succeed.

References

1. G. Herzberg; "Infrared and Raman Spectra of Polyatomic molecules" (Van Nostrand, N. York) 1945; E.B. Wilson, Jr., J.C. Decius, P.C. Cross; "Molecular Vibrations" (McGraw-Hill, N. York) 1955; H.C. Allen, Jr., P.C. Cross, "Molecular Vib-Rotors" (Wiley, N. York) 1963.

2. L.H. Little; "Infrared Spectra of Adsorbed Species" (Academic Press, N. York) 1966; M.L. Hair; "Infrared Spectroscopy in Surface Chemistry" (Marcel Dekker, N. York) 1967.

3. S.Y. Tong; Progress in Surface Sci., 7, 1 (1975).

4. R.F. Willis in "Chemistry and Chemical Engineering of Catalytic Processes" eds. G.C.A. Schuit, R. Prins, (NATO Advanced Study Institute Series, No. 39 Plenum, N. York) in press, 1980.

5.   J. Pritchard; Chap.8, this volume.

6.   D.M. Newns; Chap.2, this volume.

7.   R.F. Willis; Chap.3, this volume.

8.   D.G. Walmsley; Chap.5, this volume.

9.   J.A. Creighton; Chap.9, this volume.

10. See "Search and Discovery", Physics Today, $\underline{33}$, 18 (1980).

11. J.W. Gadzuk, H. Metiu; in "Proceedings 4th Int. Conf. on Solid Surfaces"
     Le Vide (Journal of French Vacuum Society) in press, 1980.

12. C.W. Wright, Cap.7, this volume.

13. B. Feuerbacher; Chap.6, this volume.

14. L. Dobrzynski; Chap.4, this volume.

15. N. Sheppard; Chap.10, this volume.

16. D.A. King; Chap.11, this volume.

17. R.N. Hammaker, S.A. Francis, R.P. Eischens; Spectrochimia Acta, $\underline{21}$,
     1295 (1965).

18. A.M. Bradshaw, F.M. Hoffmann; Surface Sci., $\underline{72}$, 513 (1978); A. Ortega,
     A.M. Bradshaw, F.M. Hoffmann; in "Proceedings 4th Int. Conf. on Solid
     Surfaces" Le Vide (Journal of French Vacuum Society) in press, 1980.

19. M. Moskovits, J.E. Hulse; Surface Sci., $\underline{78}$, 397 (1978).

2. Theory of Dipole Electron Scattering from Adsorbates

D. M. Newns[1]

With 11 Figures

Electron energy loss spectroscopy (EELS) is a relatively new tool for looking at adsorbate vibrations, but it is one which is rapidly expanding in popularity. The probe in this approach is a monoenergetic electron beam of energy E, where E is a few electron volts only. The beam is projected at the adsorbate covered surface at angle of incidence α, and the reflected beam is collected by an energy analyser whose resolution might be 2-10 meV, (see Fig. 1). The analyser may be fixed to look at the specular beam (early apparatus) or movable (later apparatus). In Fig. 2 is shown a typical spectrum, obtained for CO on Cu(100) at specular collection angle. One sees a very strong unscattered peak at the original beam energy of 1.3 eV, and two loss peaks at 43 meV and 260 meV below the beam energy. These losses are due to *excitation* of adsorbate vibrations at these energies, the excitation appearing as an energy *loss* of the outgoing electron. The 260 meV loss is interpreted as a C-O stretch mode, and the 43 meV loss as a mode where the CO centre of mass vibrates against the surface (loosely called metal-carbon stretch).

The physical model of the vibrational excitation of the bond by the moving electron which we discuss in this paper is that the excitation proceeds by interaction between the long range electrostatic potential of the electron and the time-varying dipole moment of the adsorbate bond in question. This "dipole-scattering" mechanism is not the only relevant one, but we shall see that it is possible to say rather clearly when it is the dominant mechanism. The dipole scattering formalism first appeared in a study of vibrational excitation of Si surfaces [2], the application to adsorbates following later [3-6].

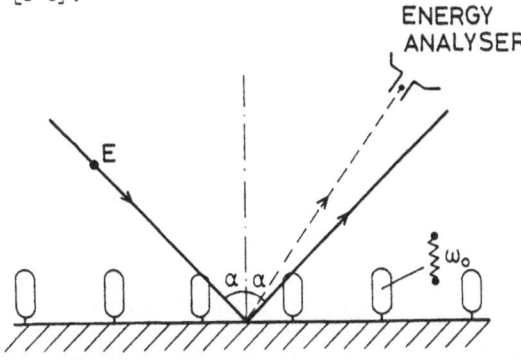

Fig. 1. Schematic EELS experimental set up; beam energy is E, angle of incidence α

[1]Permanent address: Department of Mathematics, Imperial College, London SW 7 England

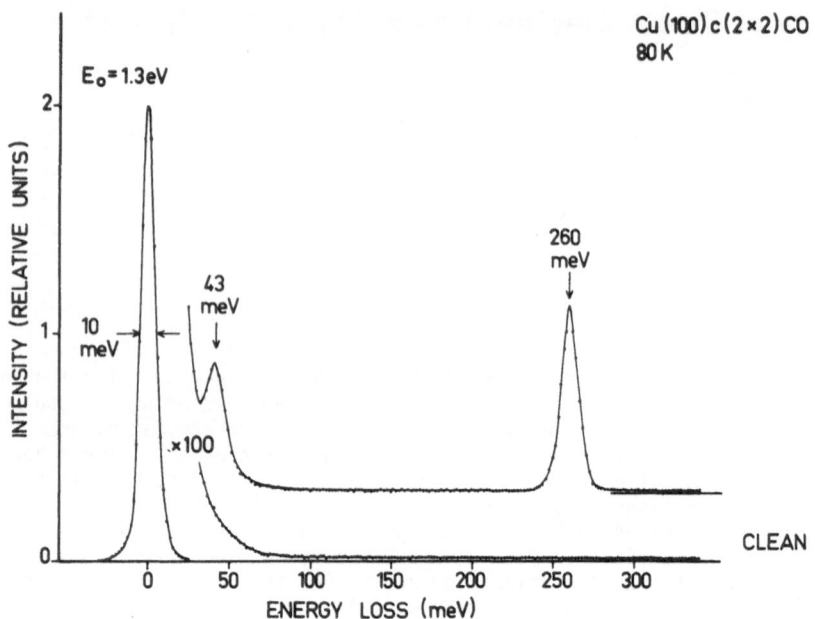

Fig. 2. Energy loss spectrum for the C(2 x 2) CO structure on Cu(100) at 80 K. E = 1.3 eV, $\alpha = 47.7^{\circ}$, analyser set to accept specular reflected beam. Taken from [1]

The theoretical approach to dipole scattering we use here is a semiclassical one [3,4], adopted because the more rigorous quantum mechanical approach [5,6] shows little significant deviation from the semiclassical results. The exciting electron is then considered to follow the classical trajectory for a reflected particle. We take the z-axis normal to surface, and introduce a capital letter notation, e.g. $\vec{X} = (x,y)$, for vectors lying in the surface plane. Taking the time of impact as $t = 0$, at the plane $z = 0$, the electron coordinate $\vec{r}_e$ may be written

$$\vec{r}_e = (\vec{X}_0 + \vec{V}t, \, v|t|), \tag{1}$$

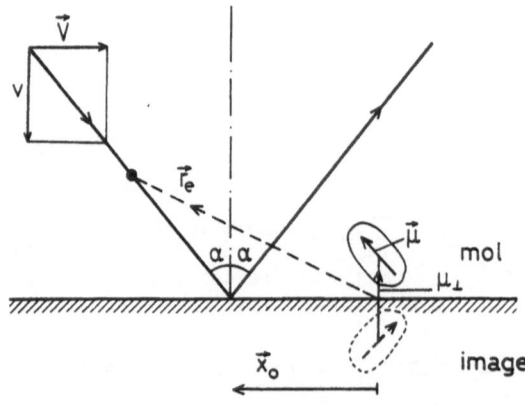

Fig. 3. Quantities entering into theory. Electron velocity $(\vec{V}, \pm v)$, electron-dipole distance $\vec{r}_e$, molecule instantaneous dipole $\vec{\mu}$ and its image in the surface, resultant normal dipole $\mu_{\perp}$

where V is the velocity parallel to the surface, v is the velocity normal to the surface, and X_0 the impact parameter in the surface plane (see Fig. 3).

Looking now at a single adsorbate molecule, the Hamiltonian H of its vibrational motion is separable into a sum

$$H = \sum_i h_i \tag{2}$$

of normal mode hamiltonians h_i, where h_i denotes the hamiltonian of the normal mode of frequency ω_i. In massless form h_i may be written

$$h_i = \frac{1}{2} p_i^2 + \frac{1}{2} \omega_i^2 q_i \tag{3}$$

where $p_i = \dot{q}_i$. The normal coordinate q_i is related by an orthogonal transformation

$$q_i = \sum_n A_{i;n,\alpha} \sqrt{m_n} x_{n,\alpha} \tag{4}$$

to the original space coordinates $x_{n,\alpha}$ (α = x,y,z) of the atoms n of mass m_n constituting the adsorbate and its neighbouring substrate atoms. The orthogonal matrix A is such as to enable the diagonal form (2) to be constructed from the original problem containing off-diagonal spring constants between different atoms n and m. For details the reader may consult standard texts on Classical Mechanics. In quantum mechanics (3) may more conveniently be written

$$h_i = \omega_i b_i^+ b_i + \frac{1}{2} \omega_i , \tag{5}$$

(we employ atomic units $\hbar = e = 1$ here and in the following), where the boson creation and destruction operators are given by

$$b_i^+ = 2^{-1/2} \left[\omega_i^{1/2} q_i - i \omega_i^{-1/2} p_i \right] \tag{6a}$$

$$b_i = 2^{-1/2} \left[\omega_i^{1/2} q_i + i \omega_i^{-1/2} p_i \right]. \tag{6b}$$

The presence of the exciting electron acts as a perturbation V on the molecule vibrational Hamiltonian (2). Within the assumptions of dipole scattering theory, the dominant electron-molecule interaction comes from the part of the electron trajectory where r_e is large (taking the molecule at the origin). Then interaction between the electronic charge and the electric field of the instantaneous molecular dipole $\vec{\mu}$ is the leading term in the interaction, which can be written

$$V = \sum_i (\nabla \frac{1}{r_e}) \cdot (\partial \vec{\mu}/\partial q_i) q_i . \tag{7}$$

At this point we make a specific assumption about the nature of the surface, that it is a low index plane of a metal. The dipole $\vec{\mu}$ then induces a screening cloud in the metal surface (see Fig. 3). Since the dipole is slowly varying at frequency ω_0, the metal electrons will follow the instantaneous dipole adiabatically. Any parallel component of $\vec{\mu}$ induces in the surface an equal and *opposite* dipole (Fig. 3). Then if r_e is large compared with the dimensions of the parallel dipole plus its screening cloud, the parallel dipole component will have negligible interaction with the electron. On the contrary the normal component μ_z of $\vec{\mu}$ is only screened out if the molecule is actually deeply embedded in the electron gas; when the molecule lies out-side the 'virtual image plane' [7] μ_z indeed induces an 'image' dipole of the same sign as itself, thus enhancing its value to $2\mu_z$. Hence on a metal sur-face only the *normal* component of $\vec{\mu}$ can interact with the electron when r_e is large. Thus (7) may be written

$$V = \sum_i \left(\frac{\partial}{\partial z_e} \frac{1}{r_e}\right) (\partial\mu_\perp/\partial q_i)q_i, \tag{8}$$

where $\mu_\perp$ is defined as the *total* normal dipole moment (including screening effects).

In some cases such as H on Ni(111) (see below) the smallness of $\mu_\perp$ suggests that H is embedded deeply enough for screening to be reducing the bond dipole moment, if indeed this can be defined independently. In other cases, such as the C-O stretch of adsorbed CO or the C-H stretch of an adsorbed hydrocarbon, it is likely that the bond dipole μ is located outside the metal screening charge, and one might think $\mu_\perp$ equal to twice μ_z. However recently it has been realised [8,9] that another mechanism exists which screens the dipole in this case, the electronic dipole polarizability of the adsorbate layer itself. For bonds well outside the surface screening charge, we then have the net screen-ing effect [8]

$$\partial\mu_\perp/\partial q_i = 2(\partial\mu_z/\partial q_i)/(1 + \alpha_e U(0)), \tag{9a}$$

where

$$U(0) = \sum_{j\neq 0} (r_j^{-3} + \tilde{r}_j^{-3}). \tag{9b}$$

In (9) α_e is the perpendicular electronic dipole polarizability of the adsorbate molecule, r_j is distance from the molecule in question at site $j = 0$ to molecule in the adsorbate layer at site j and $\tilde{r}_j$ is distance from molecule at 0 to the image of molecule j in the surface image plane.

Eq. (8) has the important consequence that only adsorbate vibrations having a component of dipole moment normal to the surface have nonzero V and can thus be active in EELS. The dipole scattering mechanism thus leads to a 'normal dipole selection rule'. A case where it might be violated is that of an adsorbate vibration located far outside the image plane on the scale of r_e (large molecule). This case has been considered by some authors [4,6].

The motion of the electron on its trajectory $r_e(t)$ leads to time depen-dence of V through (8) which becomes V(t). Let us now consider a particular normal mode of frequency ω_0, with ground state $|0\rangle$ and first excited state $|1\rangle$ (we drop subscripts i when considering just this mode). If at $t = -\infty$ the mode is in $|0\rangle$, the probability p_1 of it being found in $|1\rangle$ at $t = +\infty$ is given in second order time dependent perturbation theory as

$$p_1 = \left|\int_{-\infty}^{\infty} e^{i\omega_0 t} \langle 0| V(t) |1\rangle dt\right|^2. \tag{10}$$

Here, from (8)

$$\langle 0| V(t) |1\rangle = \gamma \, \partial |1/r_e(t)|/\partial z_e, \tag{11}$$

where

$$\gamma = \langle 0| q \, \partial \mu_\perp /\partial q |1\rangle. \tag{12}$$

γ has the dimension of a dipole moment; it is termed the *dynamic dipole moment*. Note that our definition of γ *includes* the effect of screening by the surface (see (9)). By adding (6a) and (6b) we may obtain γ in the form

$$\gamma = \partial \mu_\perp /\partial q \, (2\omega_0)^{-1/2} \langle 0| b^+ + b |1\rangle = (2\omega_0)^{-1/2} \partial \mu_\perp /\partial q. \tag{13}$$

where $\partial \mu_\perp /\partial q$ is an *effective charge*.

At this point the calculation becomes rather technical. Eq. (10) requires that we find the Fourier transform of (11). To do this we first write $1/r_e$, where $\vec{r}_e = (\vec{X},z)$, as a Fourier transform

$$\frac{1}{r_e} = \frac{1}{2\pi^2} \int d^2Q \int dq \, \frac{e^{-i\vec{Q}.\vec{X}} \, e^{-iqz}}{Q^2 + q^2} , \tag{14}$$

(14) being just the well-known Fourier transform of the Coulomb interaction. Integrating on q, we obtain the mixed representation

$$\frac{1}{r_e} = \frac{1}{2\pi} \int d^2Q \, e^{-i \, \vec{Q}.\vec{X}} \, e^{-Q|z|} Q^{-1}. \tag{15}$$

Whence we have

$$\frac{\partial}{\partial z_e} \frac{1}{r_e} = -\frac{1}{2\pi} \int d^2Q \, e^{-i\vec{Q}.\vec{X}} \, e^{-Q|z|} = -\frac{1}{2\pi} \int d^2Q \, e^{-i\vec{Q}.(\vec{X}_0+\vec{V}t)} e^{-Qv|t|} \tag{16}$$

where we have used (1). If we write (16) as a portion valid for t>0 and a portion valid for t<0 it is easy to do their Fourier transforms on time separately, then combine the results to obtain

$$\int_{-\infty}^{\infty} dt \, e^{i\omega_0 t} \langle 0| V(t) |1\rangle = - \int e^{-iQ.X_0} \, \frac{\pi^{-1} Q \, v \, \gamma}{\Omega_Q^2 + Q^2 v^2} \, d^2Q \tag{17}$$

where

$$\Omega_Q = \omega_0 - \vec{Q}.\vec{V}.$$

Substitution of (17) into (10) now yields the excitation probability p_1 for a particular scattering length $\vec{X}_0$. The measurable quantity is however the *cross-section* σ for one-phonon excitation, obtained by integrating (10) over $\vec{X}_0$. This is trivially done to yield the expression

$$\sigma = \gamma^2 \int d^2Q \, \frac{4\Omega^2 v^2}{[\Omega_Q^2 + Q^2 v^2]^2} . \tag{18}$$

11

In the spirit of the semiclassical approximation we now make use of the fact that *parallel* wave vector $\vec{Q}$ is conserved quantity. This enables us to infer that (18) not only gives the cross section for all $\vec{Q}$ but also for *each* $\vec{Q}$. We may remove the integral in (18) to obtain the differential cross section

$$d\sigma = \gamma^2 \frac{4Q^2v^2}{[\Omega_Q^2 + Q^2v^2]^2} d^2Q. \qquad (19)$$

To get a feeling for (19), let us put V=0 (normally incident electron). Then (19) becomes

$$d\sigma = \gamma^2 \frac{4Q^2v^2}{[\omega_0^2 + Q^2v^2]^2} d^2Q. \qquad (20)$$

This function is strongly peaked near the characteristic wave vector $Q_0 \sim \omega_0/v$. Suppose we take ω_0 = .01 a.u. = 270 meV (1 a.u. = 27.2 eV), and a beam energy E = .05 a.u. = 1.4 eV, then $Q_0 \approx$ .03 a.u.; the resulting Q_0 would be even smaller for a smaller ω_0 or larger beam energy. Since only small values of wavevector Q are important in the scattering we may deduce that the effective range of the electron-molecule scattering is long, of order Q_0^{-1} or at least of order 30 a.u. This is the underlying justification for using the long range dipole scattering interaction (7). These comments would not be altered qualitatively by considering the $\vec{V} \neq 0$ case.

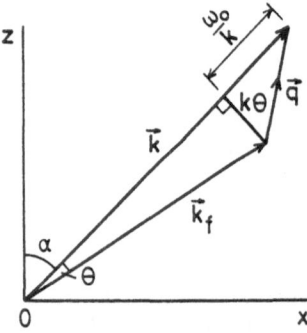

Fig. 4. Decomposition of momentum change $\vec{q}$ of electron into components perpendicular and parallel to reflected specular beam (small Θ and ω_0 assumed).

More convenient than (19) is an expression for the differential cross section in terms of measurable angles (see Fig. 4). If we define $\vec{k}$ as the wavevector of the unscattered reflected electron, $\vec{k} = (\vec{V},v) = k$ (sin α, 0, cos α) and $\vec{k}_f$ as the wavevector following excitation of a phonon, then the full scattering vector is $\vec{q} = \vec{k} - \vec{k}_f$. $\vec{q}$ has the components $\vec{q} = (\vec{Q},q_z)$. We define Θ as the angle $\vec{k}_f$ makes with the outgoing elastic beam $\vec{k}$. Specification of the azimuth ψ, the angle by which the $\vec{k}_f$, $\vec{k}$ plane is rotated about $\vec{k}$ out of the OXZ plane, (defining ψ = 0 when $\vec{k}_f$ lies on the surface side of $\vec{k}$ as in Fig. 4) fully defines the scattered beam direction $\vec{k}_f$. As discussed in the previous paragraph, we are only interested in small scattering wavevectors, which means that q/k and Θ are small. We have from energy conservation

$$\omega_0 = \frac{1}{2} (k^2 - k_f^2) \approx k(k - k_f),$$

i.e.

$$(k - k_f) \approx \omega_0/k \qquad (21)$$

and

$$k_f \sin \Theta \simeq k \Theta. \tag{22}$$

We can then label the scattering triangle as in Fig. 4. We can write down the Cartesian components of q for the normal incidence case $\alpha = 0$ and for $\psi = 0$; They are

$$\vec{q} \simeq k(-\Theta, 0, \Theta_0) \tag{23}$$

where the characteristic angle Θ_0 is given by

$$\Theta_0 = \omega_0/2E. \tag{24}$$

If ψ is nonzero $\vec{q}$ becomes

$$\vec{q} = k(-\Theta\cos\psi, \Theta\sin\psi, \Theta_0). \tag{25}$$

Finally, a rotation by angle α about OY gives the general result

$$\vec{q} = k(-\Theta\cos\alpha\cos\psi + \Theta_0\sin\alpha, \Theta\sin\psi, \Theta\cos\psi\sin\alpha + \Theta_0\cos\alpha) \tag{26}$$

so that $\vec{Q}$, the component of $\vec{q}$ parallel to the surface is

$$\vec{Q} = k(-\Theta\cos\alpha\cos\psi + \Theta_0\sin\alpha, \Theta\sin\psi). \tag{27}$$

Substituting these results into (19), we have

$$Q^2 = k^2 \left[(\Theta\cos\alpha\cos\psi - \Theta_0\sin\alpha)^2 + \Theta^2\sin^2\psi\right] = k^2\cos^2\alpha f(\Theta,\psi), \tag{28}$$

where

$$f(\Theta,\psi) = (\Theta\cos\psi - \Theta_0\tan\alpha)^2 + \Theta^2\sin^2\psi\sec^2\alpha \tag{29}$$

The function (29) first appeared in the theory of electron reflection surface plasmon excitation [10]. The denominator of (19) becomes greatly simplified

$$\Omega_Q^2 + Q^2v^2 = k^4\cos^2\alpha(\Theta^2 + \Theta_0^2). \tag{30}$$

The quantity d^2Q is given by (J = Jacobian)

$$d^2Q = J \begin{vmatrix} Q_x, & Q_y \\ \Theta, & \psi \end{vmatrix} d\Theta \, d\psi = k^2\Theta\cos\alpha \, d\Theta \, d\psi. \tag{31}$$

Collecting these results the differential cross section becomes[1]

---

[1] The notation is the same as in [3] Eq. (3) with $\lambda = \gamma$, c = k. The present (32) corrects a misprint in (3).

$$d\sigma = \frac{4\gamma^2 \cos\alpha \; f(\Theta,\psi) \; \Theta \; d\Theta \; d\psi}{k^2(\theta^2 + \theta_0^2)^2}.$$ (32)

Eq. (32), valid for general α, is almost as simple as (20). The denominator ensures that (32) peaks very strongly in the region $\Theta \lesssim \theta_0$, where θ_0 is given by (24). If again we take $E = 1.4$ eV, $\omega_0 = 270$ meV, then $\theta_0 = 0.1$ rad; we have very strong *forward scattering*. This forward scattering is itself just a consequence of the small Ω involved (see discussion following (20)). The forward scattering lobe is not actually symmetric with respects to θ. If we put $\psi = 0$, i.e. confine ourselves to the scattering plane, $f(\theta, 0)$ has a zero at $\theta = \theta_0 \tan\alpha$ (see (29)), i.e. on the surface side of the outgoing beam; this actually corresponds to the condition $\Omega = 0$ at which the intensity is seen from (19) to be zero by inspection. The differential cross section in the scattering plane for $\alpha = 45^0$, $\theta_0 = 0.1$ is sketched as a polar plot in Fig. 5. The forward scattering lobe is very pronounced. Also the asymmetry due to the node at $\theta = \theta_0$ is so pronounced as almost to confine the lobe to the normal side (negative θ) of the outgoing specular beam.

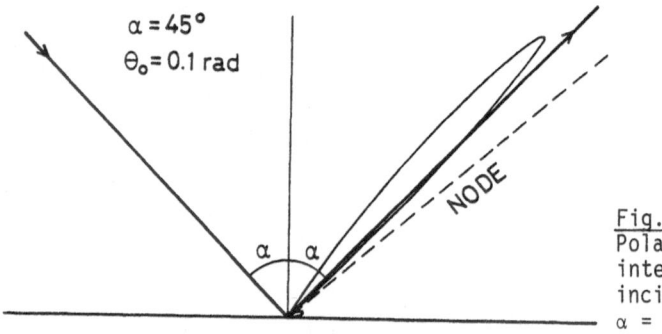

α = 45°
θₒ = 0.1 rad

NODE

Fig. 5
Polar plot of scattering
intensity in plane of
incidence for
$\alpha = 45^o$, $\theta_0 = 0.1$ rad [5].

Eq. (32) still does not take into account the aperture of the analyser. Suppose we assume that the analyser is centred on the outgoing elastic beam, and that its aperture is a circular hole, such that the analyser accepts scattered electrons lying in a cone of apex the point of reflection and semi-vertical angle θ_1, its axis being the outgoing specular beam. Then (32) may be integrated analytically to obtain

$$\sigma = \frac{\pi\gamma^2}{E} \cos\alpha \; [(t^2 - 2) \; Y + (t^2 + 2) \; \ln X] ,$$ (33)

where

$$Y = \theta_1^2/(\theta_1^2 + \theta_0^2).$$ (34)

and

$$X = 1 + \theta_1^2/\theta_0^2$$ (35)

and $t = \tan\alpha$. Note that (33) formally diverges if we allow the aperture θ_1 to go to infinity, but this is not important because it is only when $|\theta| \lesssim 3\theta_0$ that we expect the dipole mechanism to be dominant anyway due to its strong forward scattering property. We also note that σ becomes very large for grazing incidence reflection, when $\alpha \to \pi/2$; in fact $\sigma \sim 1/\cos\alpha$, just as in grazing incidence excitation of surface plasmons [8]. If the surface coverage of adsorbate is n molecules per unit area, the ratio of one phonon loss current I_1 to specular reflection current I_0 is for the specified analyser conditions

$$I_1/I_0 = \frac{\pi\gamma^2 n}{E} \cos\alpha \; [(t^2 - 2) \; Y + (t^2 + 2) \; \ln X \,]. \tag{36}$$

All the above calculations have been done for scattering of an electron from a single adsorbed molecule. The scattering intensity for a layer has then been obtained in (36) by adding the scattering intensities of each molecule. It may be enquired whether different results could be obtained by considering the coherent scattering of an electron from the Bloch phonons of an ordered layer of adsorbate. This approach was in fact used in [2] and has been also tried by the author, who found that provided coupling between the vibrations of different molecules could be neglected, identically the above results were retrieved.

The main predictions of dipole scattering theory may then be summarised as

i) There is pronounced forward scattering (Fig. 5).
ii) Only vibrations having a normal dipole moment $\mu_\perp$ varying to first order with vibrational coordinate q should be observed, a result particular to the case of molecules close above a metal surface [4,6].
iii) The scattering intensity should be related to the dynamic dipole moment γ (see (36)).

The forward scattering prediction is compared with some experimental measurements for CO on Cu(001) [1] in Figs. 6 and 7 respectively for the two modes illustrated in Fig. 2. The measurements were actually taken by rocking the specimen while keeping the electron monochromator and analyser fixed. Both experiments show very strong forward scattering. The theoretical curves illustrated are obtained by integrating numerically (32) over the conical aperture acceptance solid angle [1]; their normalization will be discussed below. It is seen that the agreement is good for the 260 meV mode. For the 43 meV mode the results are less conclusive because the angular width of the forward scattering lobe θ_0 = .022 rad is now small compared with analyser aperture θ_1 = .052 rad, whereas for the 260 meV mode at E = 1.3 eV, θ_0 = .1 rad. Other workers have also seen the strong forward scattering phenomenon, for example in the 250 meV CO vibration and in the 324 meV C-H stretch vibration of C_6H_{12}, both on Pt(111) [11]. Conclusive experiments have also been done for H on W(100) [12], discussed further below.

It is also interesting to look at the energy dependence of I_1/I_0 for the specular direction, given by (36)[2]. The measurements for CO on Cu(100), obtained for the same aperture θ_1 as in the foregoing ones of Figs. 6 and 7, are illustrated in Fig. 8 together with the theory (36), suitably normalised. The general trend is in agreement with theory though there is evidence of an unexpected structure near threshold in Fig. 8a.

The application of the dipole selection rule test (ii) requires knowledge of the adsorbate geometry. For CO on Cu(100) ARPS [13] and LEED [14] both incicate that the CO axis lies normal to the surface. This implies that C-O

_____

[2]In fact the theoretical curve differs slightly from (36) due to avoiding the small $-\omega_0$ approximation of Fig. 4.

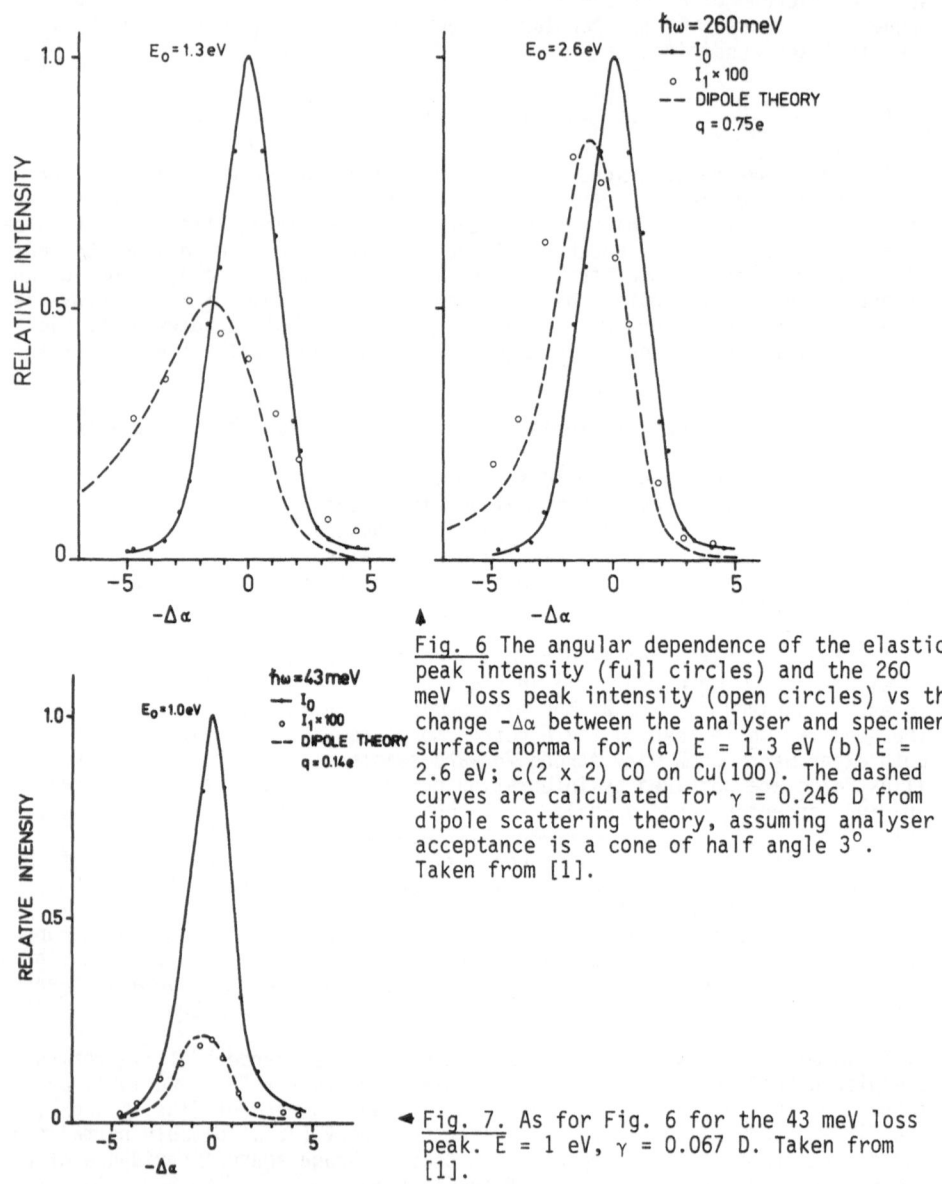

Fig. 6 The angular dependence of the elastic peak intensity (full circles) and the 260 meV loss peak intensity (open circles) vs the change $-\Delta\alpha$ between the analyser and specimen surface normal for (a) E = 1.3 eV (b) E = 2.6 eV; c(2 x 2) CO on Cu(100). The dashed curves are calculated for γ = 0.246 D from dipole scattering theory, assuming analyser acceptance is a cone of half angle 3°. Taken from [1].

Fig. 7. As for Fig. 6 for the 43 meV loss peak. E = 1 eV, γ = 0.067 D. Taken from [1].

stretch and CO-Cu stretch but *not* the CO-Cu bending vibration should be seen. One mode (260 meV) unmistakeably assignable to C-O stretch is seen. The 43 meV is then assigned to CO-Cu stretch; the absence of any third mode, even in recent high-resolution measurements is evidence for the operation of the dipole selection rule. Further elegant evidence for the dipole selection

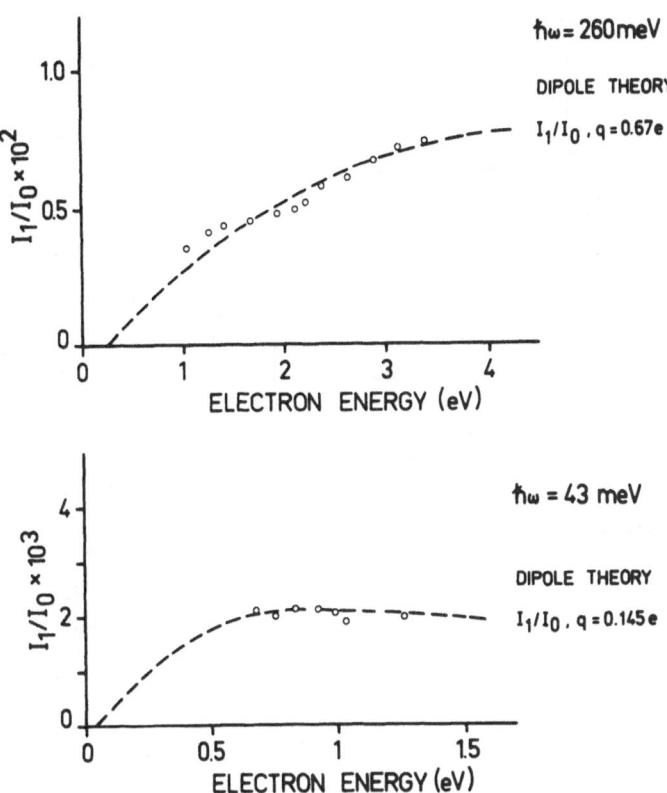

Fig. 8 Relative loss intensity, (I_1/I_0) vs primary electron energy E of the
(a) C-O and (b) Cu-CO stretching modes for C(2 x 2) CO on Cu(100).
Full circles, experimental data, dashed curve dipole scattering
theory with (a) γ = .22 D, (b) γ = .07 D. Taken from [1].

rules comes from work on the β_1 H on W(100) system [12]. Fig. 9a shows an
EELS spectrum taken at the specular direction, showing only a 130 meV loss.
The angle dependence of this loss is shown in Fig. 10, and shows typical forward scattering behaviour. Interestingly, the Ref. [12] work also took spectra
well away from the specular direction (Fig. 9b). In this region additional
losses are seen at 80, 160 and 260 meV, which are discussed further in the
article in this volume by R.F. Willis. The 80 and 160 meV losses are assigned
to non-dipole active vibrations of H in a bridge-bonded configuration, and
the 260 meV loss to a two-phonon satellite on the 130 meV vibration [12].
This work shows that bridge bonded H on W(100) has three modes; the dipole
active mode shows typical dipole-scattering behaviour, while the other two
modes and a two-phonon excitation do *not* show such behaviour but can be
observed at large angles. Two phonon excitation is expected to be a very weak
process in the dipole scattering mechanism, but not necessarily in an "impact"
mechanism; we refer the reader to a more complete discussion [16].

Dipole scattering from parallel vibrations becomes possible, it should be
noted, if the molecule is too far above the metal surface [4,6] or if the
conducting surface is replaced by a dielectric. One then expects to see
narrow scattering lobes from parallel vibrations [6]. The scattering is however

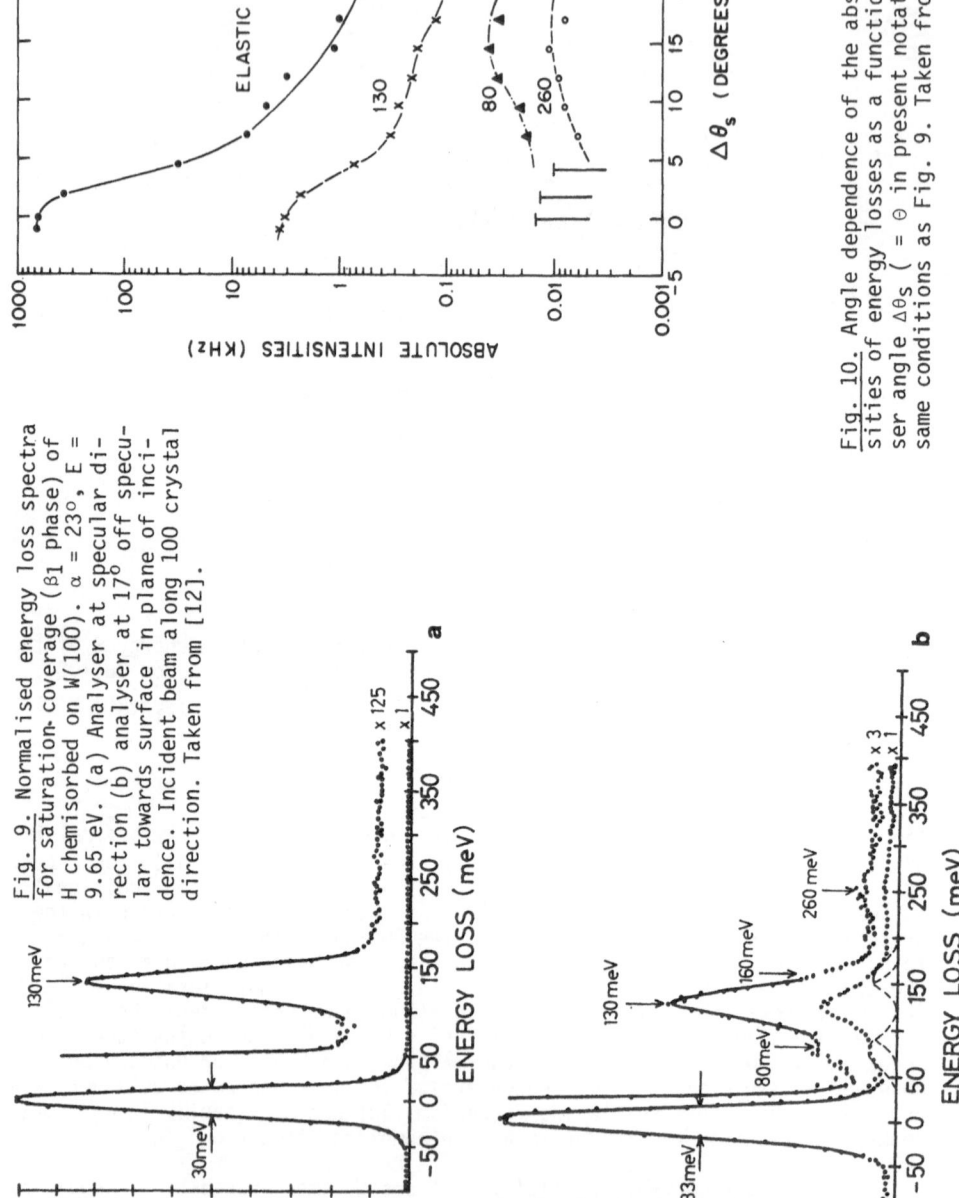

Fig. 9. Normalised energy loss spectra for saturation-coverage (β1 phase) of H chemisorbed on W(100). α = 23°, E = 9.65 eV. (a) Analyser at specular direction (b) analyser at 17° off specular towards surface in plane of incidence. Incident beam along 100 crystal direction. Taken from [12].

Fig. 10. Angle dependence of the absolute intensities of energy losses as a function of analyser angle Δθs (= Θ in present notation, ψ = 0), same conditions as Fig. 9. Taken from [12].

zero in the plane of incidence for vibrations normal to the plane of incidence; for vibrations parallel to the surface, but in the plane of incidence,
it has a node at $\theta = \theta_0 \tan \alpha$ [6]; while the node in the normal vibration
scattering no longer lies exactly at $\theta = \theta_0 \tan \alpha$ [6]. Actually these two
conditions or selection rules on parallel vibrations are quite general, i.e.
independent of scattering mechanism, as has recently been demonstrated [15],
so that scattering from parallel vibrations is never seen along the $Q = 0$
direction.

The relationship of the scattering intensity to the dynamic dipole moment
γ, defined by (8), now needs to be investigated. We recall that γ is defined
to include any screening or image effects at the surface, and it does not
depend on how the Lagrange coordinates are defined; γ is thus directly related to experimental measurements. Fortunately another measurement, IR reflection spectroscopy, also exists which measures the same quantity γ. The determination of γ from experimental IR intensities has been discussed by Persson
[17]. In Table I we compare γ values[3] for the CO stretch vibration of adsorbed
CO derived in [17] from IR measurements with γ values from EELS. These were
determined from the normalization of the curves in Figs. 6 and 8 for CO on
CU(100), and from normalization of the Fig. 8 plot [16] for CO on Ni(100).

Table I γ for adsorbed CO stretch

Method	System	γ (Debye)	Ref
IR	Cu(100)	.20	[18]
	Pt(111)	.184	[19]
EELS	c(2 x 2) on Cu(100)	.22, .24	[10]
	c(2 x 2) on Ni(100)	.20	[16]

It is seen from Table I that the agreement between the IR-derived and EELS-
derived values of γ is very satisfactory. However caution regarding the former
is indicated, because of the lack of coverage measurement in the work of [18]
and [19]; the relationship between absorption intensity and coverage is non-
trivial, because of screening by the electronic polarizability of the adsorbed
layer (see (9)). The reader should also be referred to the early work of
Ibach [20] on comparisons between IR and EELS cross sections.

In the case of the C-O stretch mode, where the active atoms are located
outside the surface for chemisorbed CO with its axis normal to the surface,
it is worth trying to disentangle a bond dynamic dipole moment
$\mu_z = <0| q \, \partial \mu_z/\partial q |1>$ which, using (9), is related to γ by

$$\mu_z = \gamma(1 + \alpha_e U(0))/2. \qquad (37)$$

For U(0) we may for rough purposes employ the Topping approximation

---

[3] If Table I is compared with the corresponding table in [17], it should be
noted that our γ is defined as twice the quantity μ of [17].

$$U(0) = 18 \, n^{3/2},\tag{38}$$

where n = coverage/unit area. If we choose the gas phase polarizability of CO parallel to the CO axis, α_e = 2.5 A^3 [21], and use (38), a rough estimate of (37) for c(2 x 2) CO on Cu is

$$\mu_z \simeq 1.0 \, \gamma.\tag{39}$$

Thus very roughly the doubling of μ_z due to its image is neutralised by the screening effect of the adsorbate layer. More precise estimates have been made in [8], with similar conclusions. Thus from Table I μ_z is about 0.22 D. This value is about right for explaining the dipole-dipole coupling between different adsorbate molecules [8], whose measurement is possible by IR experiment on isotopically mixed layers [8,22]. This value is however much larger than the gas phase value μ_z = .104 D [23], the extra contribution for adsorbed CO being attributed [8,9] to electron transfer from the substrate Fermi level into the empty antibonding $2\pi^*$ orbital of gas phase CO. With increasing CO distance, the $2\pi^*$ orbital should go down in energy enabling it to pick up more charge, which together with its image combines to form a dipole of the same sign, C^+O^-, as obtained by stretching the isolated CO molecule [24]. This hypothesis is lent plausibility by the observation of correspondingly large μ_z values on carbonyls, for example μ_z = .24 D on Ni(CO)$_4$ [8,25]. Furthermore, the μ_z value deduced for molecularly adsorbed N_2 with axis normal to the surface is found to be μ_z = .08 D [9], which is similar to the *difference* .22 - .1 $\simeq$ .12 between adsorbed and gas phase CO, as would be expected since *all* the μ_z for the symmetric molecule N_2 must come from the charge transfer mechanism. In this context, we note that the existence of some occupation of the $2\pi^*$ orbital for chemisorbed CO (and carbonyl CO) is the widely accepted explanation for the lower C-O stretch frequencies of adsorbed CO relative to the gas phase; for example 255 meV for c(2 x 2) CO on Ni (100) [16], 260 meV for c(2 x 2) CO on Cu(100) [1], compared with 266 meV in gas phase CO [26].

Thus developments in deconvolution of a bond dynamic dipole moment μ_z from the measured quantity γ have recently led [8,9] to a larger value of μ_z which seems plausible, but the difficulties inherent in this procedure make this an area of EELS theory likely to be subject to further developments.

All the theory given so far has been based on the rectilinear trajectory (1), which neglects the image force between the electron and the surface. Consider the ratio r between the image energy 1/4d at the characteristic distance for phonon excitation d = $1/(\Theta_0 k)$, and the perpendicular component of kinetic energy $k^2 \cos^2\alpha/2$,

$$r \simeq \omega_0/(4k^3 \cos^2\alpha).\tag{40}$$

r can be large at sufficiently grazing incidence or when the electron has low kinetic energy. The former case is found in inelastic plasmon scattering at extremely grazing incidence where the effect has been estimated and compared with experiment [27]; the trajectory can be regarded as bending in towards the surface, so that the perpendicular velocity can never be less than the quantity k = $(2d)^{-1/2}$. This effect ultimately limits the apparent divergence in cross section (33) as one goes to grazing incidence [27]. The analogous effect in vibrational EELS spectroscopy was worked out independantly and

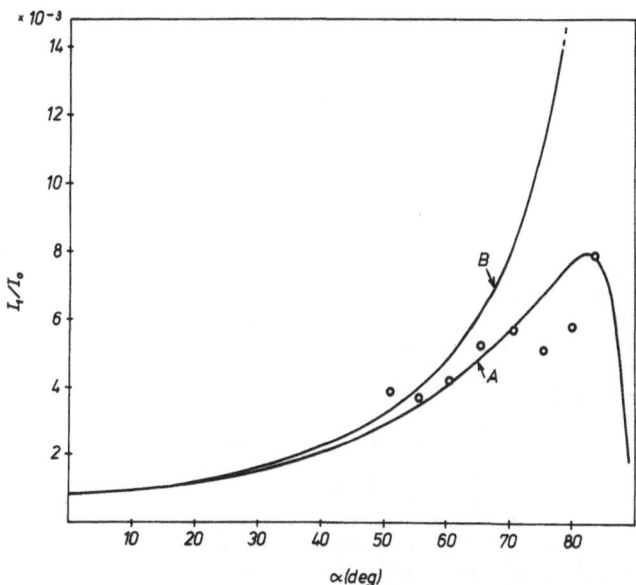

Fig. 11   Relative loss intensity for the 237 meV C-O stretching vibration for
c(4 x 2) CO on Pt(111), as a function of α at E = 5 eV. Analyser
centred on specular reflected beam. Open circles, experimental data.
Curves A and B are theoretical curves, for γ = .3 D, calculated respectively with and without including the image force. Taken from [29].

rather thoroughly by Persson [28] using however the quantum mechanical treatment of the electron motion. The vibrational EELS case turns out to be a favourable one for observation of the effect because k is small in (40). A
comparison with experiment [29] is illustrated in Fig. 11, where although
the data are rather noisy the non-divergence of cross section at large α is
clearly seen.

In conclusion, the most important consequence of dipole scattering theory
is the concentration of inelastic differential cross section into a narrow
lobe, of width controlled by the characteristic angle θ_0, lying just above
the outgoing specular beam. The origin of this forward scattering phenomenon
is the 'soft' or extended target presented by the molecule's fluctuating dipole field to the electron, which sees a spatial extension $\sim (k\theta_0)^{-1}$. θ_0 is
in practice small and the extension large, the momentum transfer $\sim k\theta_0$ small.
Provided the dynamic dipole moment γ is reasonable, scattering measured near
the lobe region should be dominantly dipole scattering and should show the
normal dipole selection rule. Measurements too far outside the lobe incur
a large momentum transfer for which many of the assumptions of dipole scattering theory break down, while the dipole scattering intensity has then
fallen off enabling other scattering mechanisms to compete. If γ is very
small, as seems to be the case for H on Ni(111), then a negligible forward
scattering lobe is seen [30].

Acknowledgements

Valuable discussions with many people, especially S. Andersson, R. Brako, H. Ibach, B. Persson, M. Persson, K.L. Sebastian and R.F. Willis have benefited this article. I am also grateful for the hospitality of the Chalmers University of Technology, financed by a Nordita visiting professorship, where this article was written.

References

1. S. Andersson, B.N.J. Persson, T. Gustafsson, E.W. Plummer: Solid State Commun. 34, 473 (1980)
2. E. Evans, D.L. Mills: Phys. Rev. B5, 4126 (1972)
3. D.M. Newns: Phys. Lett. 60A, 461 (1977)
4. F. Delanaye, A. Lucas, G.D. Mahan: Surf. Sci. 70, 629 (1978)
5. B.N.J. Persson: Solid State Commun. 24, 573 (1977)
6. D. Sokcevic, Z. Lenac, R. Brako, M. Sunjic: Z. Physik B 28, 273 (1977); Z. Lenac, M. Sunjic, D. Sokcevic, R. Brako: Surf. Sci 80, 602 (1979)
7. N.D. Lang, W. Kohn: Phys. Rev. B7, 3541 (1973)
8. B.N.J. Persson, R. Ryberg: Solid State Commun., in press
9. B.N.J. Persson, M. Persson: Solid State Commun., in press
10. E.A. Stern, R.A. Ferrell: Phys. Rev. 120, 130 (1960)
11. H. Ibach: private commun.
12. W. Ho, R.F. Willis, E.W. Plummer: Phys. Rev. Lett. 40, 1463 (1978)
13. C. Allyn, T. Gustafsson, E.W. Plummer: Solid State Commun. 24, 531 (1977)
14. S. Andersson, J.B. Pendry: Phys. Rev. Lett. 43, 363 (1979)
15. S.Y. Tong, C.H. Li, D.L. Mills: Phys. Rev. Lett. 44, 407 (1980); K.L. Sebastian: Phys. Rev. Lett. 44, 1414 (1980)
16. S. Andersson, J.W. Davenport: Solid State Commun. 28, 667 (1978)
17. B.N.J. Persson: Solid State Commun. 30, 163 (1979)
18. K. Horn, J. Pritchard: Surf. Sci. 55, 701 (1976)
19. H.J. Krebs, H. Lüth: Appl. Phys. 14, 337 (1977)
20. H. Ibach, Surf. Sci. 66, 56 (1977)
21. Landolt-Börnstein, Band 1, 3. Teil (Springer Berlin, Heidelberg, New York 1951): p. 510
22. P. Hollins, J. Pritchard: Surf. Sci. 89, 486 (1979)
23. R.A. Toth, R.H. Hunt, E.K. Plyler: J. Molec. Spectrosc. 32, 85 (1969)
24. O. Gunnarsson, J. Harris, R.O. Jones: J. Chem. Phys. 67, 3970 (1977); 68, 1190 (1978)
25. M. Bigorne: Spectrochem. Acta 32 A, 673 (1976)
26. G. Herzberg: *Infrared and Raman Spectra* (van Nostrand: Princeton, NJ 1945)
27. J.P. Muscat, D.M. Newns: Surf. Sci 64, 641 (1977)
28. B.N.J. Persson: Surf. Sci. 92, 265 (1980)
29. B.N.J. Persson, H. Ibach: Surf. Sci., in press
30. E.W. Plummer, W. Ho, S. Andersson: private commun.

3. Angle and Energy Dependent
Electron Impact Vibrational Excitation of Adsorbates

Roy F. Willis

With 23 Figures

3.1. Background

In a previous chapter in this volume, NEWNS [1] has outlined the electron
scattering conditions under which image dipole theory is generally applicable, viz: scattering into a narrow angle about the specularly reflected
beam direction produces electron energy losses arising from the vibrational
excitation of modes that possess a net change in dipole moment perpendicular to a metal surface. In this chapter, we will be concerned with the
limitations of this theory. In particular, we will investigate circumstances
which lead to a break down in this "surface dipole selection rule" to give
wide angle scattering from adsorbate modes vibrating parallel, as well as
perpendicular, to the surface. The importance of this wide angle scattering
(due to the short range part of the electron scattering potential) is that
it provides detailed information on the structure of the surface adsorbate
system. The number of vibrational modes, as revealed by the angle dependence
of the differential inelastic scattering cross sections, relates to the point
group symmetry of the adsorbate's site and geometry [2]. That is, the cross
section for any given mode should serve to distinguish between scattering
from modes with displacements parallel as opposed to perpendicular to the
surface plane. The impact energy dependence of the cross sections provides
information concerning the scattering mechanism, as well as the adsorbate
surface structure.

Our objective, therefore, is to delineate what can be learned about the
vibrational modes of adsorbate layers from a close examination of impact
energy and angle dependent electron-energy-loss spectroscopy (EELS) studies.
There have been relatively few such studies. Much of the work to date is concentrated mainly on the identification of adsorbed species within the context
of the dipole scattering configuration; these studies have been extensively
reviewed [3]. In what follows, we will limit our discussion to an overview
of various electron scattering theories which provide us with a deeper in-
sight into the vibrational excitation mechanisms.

We begin (section 2) by making a clear distinction between electron scattering by the long range (dipole) as opposed to the short range (impact) part
of the overall scattering potential of an adsorbed molecule. We approach
the problem by first considering the hypothetical case of low energy electron
scattering from an isolated, oscillating dipolar molecule (such as CO) oriented with its dipole fixed in space [4,5]. The theory of inelastic electron
scattering from molecules in the gas phase is well established [6] and our
objective is to see if the excitation mechanisms applicable to the free

molecule carry over to the adsorbed phase. This allows us to explore more clearly the specific role imposed by scattering from the substrate surface itself. In section 3, we restate surface dipole theory in the context of scattering from the perpendicular and parallel modes of a single adsorbate molecule [7]. The importance of the electron reflectivity and phase shift is underlined. In section 4, we extend the discussion to the long range interaction with a surface lattice array of oscillating dipoles, couched in terms of one-phonon excitation surface dipole theory [8]. The parallel modes are revealed as a consequence of short range quadrupole scattering [9]. A more comprehensive multiple scattering theory of one-phonon excitation is outlined in section 5 [10]. Finally, results for scattering from the hydrogen-induced modes on W(100) are discussed in terms of the predictions of the various theories, section 6. Concluding remarks are made in section 7.

3.2 Inelastic Electron Scattering from Oriented Molecules

In general the scattering of an electron from a molecule with N electrons can be described by the Hamiltonian

$$H = H_M + T_e + V(\vec{r}, \vec{R}) \quad . \tag{1}$$

H_M represents the target molecule, T_e is the kinetic energy of the scattering electron and $V(\vec{r}, \vec{R})$ is the interaction potential between the scattering electron and the target molecule. Since the interaction time is usually small ($\sim 10^{-16}$ sec) compared with a vibrational period ($\sim 10^{-10}$ sec), vibrational excitation can be treated in the adiabatic-nuclei approximation where the Born-Oppenheimer separation into electronic and nuclear wavefunctions is applicable. That is, the wavefunction for the full Hamiltonian (1) is given by

$$\psi(\vec{r}, \vec{R}) = \psi_M(\vec{x}, \vec{R}) \, \psi_e(\vec{r}, \vec{R}) \quad ; \tag{2}$$

$\psi_e(\vec{r}, \vec{R})$ is the scattering orbital for fixed nuclear coordinates $\vec{R}$ and is a solution of $H_e = T_e + V(\vec{r}, \vec{R})$. The target wavefunction $\psi_M(\vec{x}, \vec{R})$ is an eigenfunction of the molecular Hamiltonian H_M which can be further separated into a product of electronic, vibrational and rotational wavefunctions

$$\psi_M(\vec{r}, \vec{R}) = \psi_{el}(\vec{x}) \, \psi_{vib}(v, J, R) \, \psi_{rot}(J, M, \hat{R}) \tag{3}$$

where v,J,M denotes the usual vibrational and rotational states of a molecule in its ground electronic state. In electron scattering experiments, the rotational transitions are not resolved and it is usual to take a thermal average of the initial rotational states and sum over all possible final rotational states [4]. In the case of molecules with fixed orientation, the rotational motion is effectively frozen out.

Vibrational excitation can occur through the interaction of the incident electron with the long range oscillating dipole potential of a target molecule or via the short range atomic potentials. The dependence of the inelastic cross sections on the impact energy and scattering angle are expected to be very different. It is instructive therefore to first compare scattering from an oriented molecule with that of an oscillating dipole, neglecting any effects due to the substrate. This was an approach first taken by DAVENPORT, HO and SCHRIEFFER [4]. In their approach, they applied the X-α multiple scattering formalism first developed by JOHNSON and co-workers [11] for bound states and subsequently extended to continuum states by DILL and DEHMER [12].

In this method, the interaction potential (1) is approximated

$$V(\vec{r},\vec{R}) \;=\; V_N(\vec{r},\vec{R}) \;+\; V_H(\vec{r}) \;+\; V_{xc}(\vec{r}) \tag{4}$$

where $V_N(\vec{r},\vec{R})$ is the nuclear attraction term, $V_H(\vec{r})$ is the static electron-electron interaction or Hartree term and $V_{xc}(\vec{r})$ is an exchange and correlation function which is assumed to be proportional to $n^{1/3}$, where n is the local electron density

$$V_{xc}(\vec{r}) \;=\; -3\alpha e^2 (3n/8\pi)^{1/3} \tag{5}$$

α being a parameter of the order 0.7.

A second approximation is that the atomic potentials are spherically and volume averaged to a "muffin-tin" form. That is, non-overlapping spheres are constructed around each nucleus and an outer sphere is placed tangentially around the entire molecule, as shown schematically for the CO molecule, Fig.1.

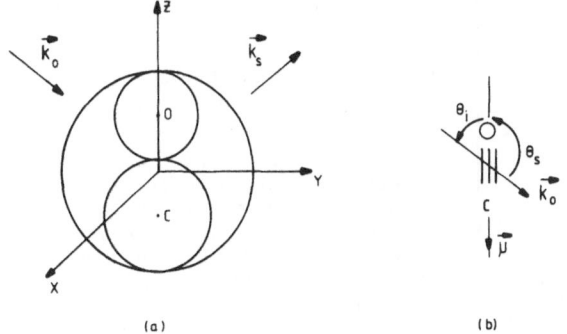

(a)   (b)

Fig.1 a) Muffin-tin spheres approximation for electron scattering ($\vec{k}_0,\vec{k}_s$) from an oriented CO molecule.
       b) Scattering geometry with respect to dipole orientation.

The potential is spherically averaged within the atomic spheres and that between the inner and outer spheres is volume averaged to a constant. At the outer sphere boundary, the X-α potential is smoothly matched into an asymptotic polarization potential ($\sim 1/r^4$). The electron density which determines the X-α potential (5) is the total density for the system, target molecule plus incident electron.

Given the Hamiltonian (1), the full transition amplitude between target states M and M' and scattering electron states $\vec{k}_0$ and $\vec{k}_s$ is given by

$$f(\vec{k}_0 M \rightarrow \vec{k}_s M') \;=\; (-m/2\pi\hbar^2) \; < e^{i\vec{k}_s\vec{r}}\psi_{M'}| \; V \; |\psi(\vec{r},\vec{R})> \tag{6}$$

25

m is the electron mass and $\psi(\vec{r},\vec{R})$ is the eigenfunction of the full Hamiltonian (1) in the adiabatic approximation (2), where the scattering orbital has the asymptotic form

$$\psi_e(\vec{r},\vec{R}) \rightarrow e^{i\vec{k}\cdot\vec{r}} + f_{\vec{k}_0\vec{k}_s}(\vec{R})\, e^{ikr}/r \tag{7}$$

The wavefunctions and their derivatives are matched at the boundaries of the muffin tin spheres and the vibrational wavefunctions of the target molecule $\psi_M(\vec{R})$ are assumed to be those of a harmonic oscillator with vibrational frequency ω_{vib}.

For electron scattering by molecules in the gas phase, the cross sections for vibrational excitation are dominated by *negative ion resonances* [6]. The incoming electron is temporarily trapped in a virtual bound state which can be viewed as an antibonding level of the molecule. These are one-electron effects in the sense that they do not involve any electronic excitation of the neutral molecule. The most pronounced consequence of such a negative ion resonance is an enhancement of the vibrational excitation cross sections at the resonant impact energy. This can be seen clearly in the multiple scattering results for oriented CO, Fig.2.

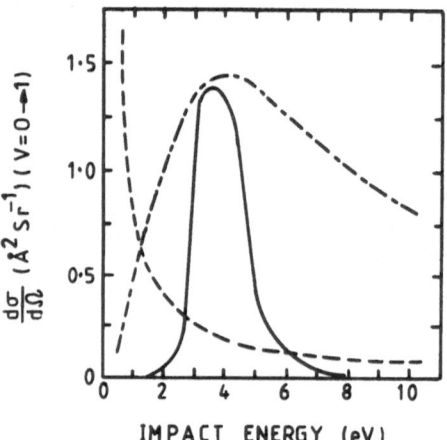

Fig.2 Impact energy dependence of the $v = 0 \rightarrow 1$ vibrational excitation cross section for an oriented CO molecule at 45° angle of incidence to dipole axis [5]. Full curve (negative ion resonance, $\theta_S = 85°$); broken curve (dipole scattering, forward direction, $\theta_S = 0°$); chain curve (dipole scattering, $\theta_S = 2°$).

The calculation yields a resonance maximum at ~ 3.75 eV impact energy [4].

This scattering due to the short range atomic potentials is very different from that for scattering from a dipole pointing in the same direction as the oriented CO. In this case, the dipole potential is simply

$$V(\vec{r},\vec{R}) = \frac{-e\mu(\vec{R})}{r^2}\cos\theta\,(\hat{r},\hat{\mu}) \tag{8}$$

where θ is the angle between the unit vectors of the spatial point $\vec{r}$ and the dipole moment $\vec{\mu}$, and $\vec{R}$ is the separation between the point diple charges (Fig.1). Assuming a plane wave representation of the scattered electron (first Born approximation), the differential scattering cross section is given by

$$\frac{d\sigma}{d\Omega} = \frac{2me}{\hbar^2}\,|<v'|\,\mu(R)\,|v>|^2\,\frac{1}{q^2}\left(1 - \frac{\sin qb}{qb}\right)\cos^2\theta'(\hat{q},\hat{\mu}) \tag{9}$$

26

where v and v' denote the initial and final vibrational states; q is the magnitude of the momentum transfer between k_0 and k_s; and θ' is the angle between the momentum transfer $\vec{q}$ and the dipole moment $\vec{\mu}$. The dipole potential cuts off at a distance corresponding to a sphere of small radius b. The results shown in Fig.2 are calculated [5] without cutting off the range of the dipole potential (b = 0) for an energy loss corresponding to the v = 0 → 1 CO vibrational transition (269 meV). No sharp resonance is found and the variation of cross section with impact energy depends on the angle of scattering, whereas that for the negative ion resonance does not.

The difference between dipole and resonance scattering is also particularly marked in the angular distribution dependence, Fig.3a. As suggested by relationship (9), the cross section for dipole scattering should peak strongly close to the forward direction where q is small. Due to the finite energy loss, maximum intensity occurs a few degrees away from the forward direction. At this angle of scattering, the impact energy dependece (Fig.2) shows a broad maximum at ∿ 4 eV impact energy. The dependence is quite different for large scattering angles θ_s ∿ 85° due to the interplay between $\vec{q}$ and $\vec{\mu}$ via the term $\cos^2\theta'$ (9).

In contrast, short range resonant scattering produces a much wider angular distribution, the actual pattern reflecting the symmetry of the π resonance state, Fig.3. In the case of CO both p- and d-wave components contribute due to the lower symmetry of the CO molecule compared with, say molecular N_2 in which the scattering is due almost entirely to the d orbitals [4]. In the case of dipole scattering, the potential is much longer range such that the scattering is not at all sensitive to the molecular charge distributions which give rise to the net dipole moment. However, as soon as the dipole potential is cut off, the scattering becomes diffuse. In Fig.3a, the radius of the cutoff was chosen [5] to be that of the outer sphere of the oriented CO molecule (∿ 1.13 Å). The intensity of this diffuse scattering is concentrated about the dipole axis.

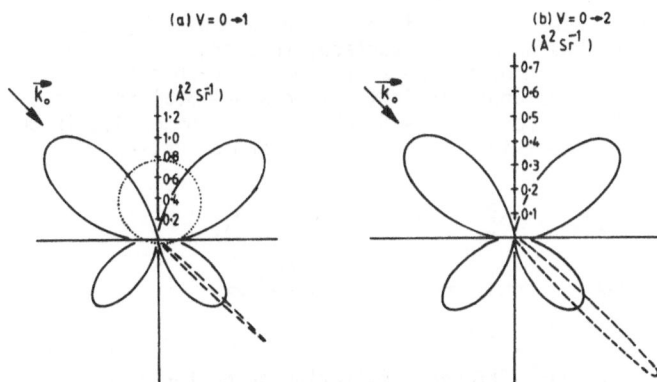

Fig.3 (a) Angle dependent scattering of the v = 0 → 1 vibrational excitation cross section for an oriented CO molecule at 45° angle of incidence and 4 ev impact energy [5]. Full curve (negative ion resonance); broken curve (infinite-range dipole scattering); dotted curve (cut-off dipole scattering, b = 1.13 Å)

Fig.3 (b) Angle dependent scattering of the v = 0 → 2 overtone vibrational excitation cross section for oriented CO. Full curve (negative ion resonance); broken curve (infinite-range dipole scattering): Note the difference in scale compared with (a).

Away from the resonant impact energy, the angular patterns are essentially preserved but with a reduction in intensity in the case of the negative ion resonance. Also, the shape of the negative ion resonance angular distributions (being a molecular property) should be the same for overtones ($v = 0 \rightarrow 2$) as for fundamentals ($v = 0 \rightarrow 1$). This is seen to be the case in Fig.3b. The only difference is that the electrons are scattered over a wider angle due to the larger energy loss. The angular distribution resembles closely the four-lobed pattern although the intensity is a factor of 2.5 lower.

In dipole scattering, the dipole moment can be expanded in a Taylor series

$$\mu(R) = \mu(R_0) + \frac{d\mu}{dR}\Big|_{R=R_0} u + \frac{1}{2!}\frac{d^2\mu}{dR^2}\Big|_{R=R_0} u^2 + \ldots \ldots \tag{10}$$

where the $u = R - R_0$ is the vibration normal coordinate for an equilibrium separation R_0 of the point dipole charges. First overtones can be excited by the u^2 term for non-vanishing $d^2\mu/dR^2|_{R=R_0}$ and are commonly observed in electron scattering with an intensity reduced typically 1/10 that of the fundamentals [2,6]. In short range impact scattering, the operator for vibrational excitation (6) cannot be separated into products that depend separately on the electronic and nuclear coordinates, as in the case of the dipole operator (9) and (10).

It can be seen, therefore, that negative ion resonance scattering dominates the impact energy and angular dependence in the gas phase. The question is to what extent will this behaviour persist when the molecule is chemisorbed on a metal surface?

3.3 Surface Dipole Scattering

In the gas phase, electron-induced dipole interaction is a weak process, whereas for a molecule adsorbed on a metal surface, this process is enhanced due to the image screening effect of the metal's electronic charge density. Numerous theoretical papers have appeared in the literature dealing with this effect [7,13,14,15,16], as reviewed by NEWNS [1]. Our objective is to explore the surface dipole effect in relation to wide angle scattering.

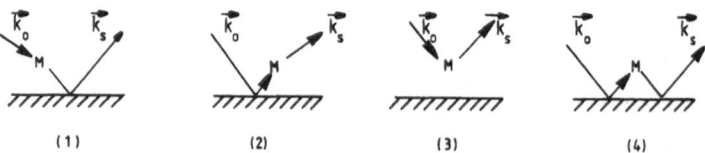

Fig.4 Schematic drawing of four different scattering paths for a point dipole (M) located above a semi-infinite metal surface [7].

Figure 4 shows schematically four different scattering paths for a point scatterer (M) located above a semi-infinite planar metal surface. The incident electron is characterized by a wave vector $\vec{k}_0$ which together with inelastic electron's wave vector $\vec{k}_s$ defines

$$\frac{\hbar^2}{2M}(k_0^2 - k_s^2) = \hbar\omega_{vib} \tag{11}$$

$\hbar\omega_{vib}$ being the vibrational loss energy. The different scattering paths (Fig.4) can be generalized as being representative of the two catagories of scattering process, viz: the long range vs the short range contributions to the vibrational interaction potential. The long range (dipole) potential leads to small angle forward scattering either followed or preceeded by elastic backscattering from the substrate (processes 1 and 2, Fig.4). In contrast, short range scattering involving the molecular adsorbate (M) (processes 3 and 4) leads to large angle distributions.

The system can be described by an Hamiltonian similar to (1)

$$H = H_m + T_e + U + V(\vec{r},\vec{R}) \tag{12}$$

where the additional term U is now the potential of the metallic surface that produces specular reflection. The electric field of the molecular vibrations is approximated by a point dipole located a distance z_0 above the metal surface. The interaction between the electrons and the vibrating dipole is via its long range dipole field

$$V(\vec{r},\vec{R}) = -e\mu(\vec{R}) \cdot \frac{\vec{r}}{r^3} \tag{13a}$$

Using the notation of LENAC et al. [7], we can expand $\mu(\vec{R})$ into

$$\mu(R) = \mu_S + u\left(\frac{d\mu}{du}\right)\hat{\mu}_0 \tag{13b}$$

where u is the vibration normal coordinate in units of the r.m.s. displacement of a harmonic oscillator in its ground state; $\hat{\mu}_0$ is a unit vector in the direction of the molecular dipole and μ_S is the static dipole moment. For this long range potential, we can write the electronic wavefunctions as independent of $\vec{R}$ [17]

$$\psi_k(\vec{r},\vec{R}) = \psi_k^{\pm}(r) \simeq e^{i\vec{k}_{//} \cdot \vec{r}}\left[e^{\pm ik_\perp z} + A(\vec{k})e^{\pm 2i\delta(\vec{k})}e^{\pm ik_\perp z}\right] \tag{14}$$

That is, the eigenstates of $(T_e + U)$ are specularly reflected plane waves, where (+,-) refer to the incoming (initial) and outgoing (final, time-reversed) states, respectively. The amplitude $A(\vec{k})$ and the phase shift $\delta(\vec{k})$ determine the reflectivity A^2 for the surface potential scattering in the specular direction. Assuming the Born-Oppenheimer approximation (2), the transition rate for the process in which the molecular dipole is excited from vibrational state $<\phi_i(R)|$ to $<\phi_f(R)|$, and at the same time one-electron scattered from state $\psi(\vec{k}_0)$ into $\psi(\vec{k}_s)$ requires that we must solve the two matrix elements [17]

$$<\phi_f| \vec{\mu} | \phi_i > \tag{15a}$$

and

$$<\psi_{\vec{k}_s} | \frac{\vec{r}}{r^3} | \psi_{k_0} > \tag{15b}$$

the vibrational transition matrix element (15a) for a harmonic oscillator taking the form

$$\left| < \phi_n | u\left(\frac{d\mu}{du}\right) | \phi_0 > \right|^2 = \frac{\hbar}{2M\omega_0}\left(\frac{d\mu}{du}\right)^2 \tag{15c}$$

with n = 1 for the singly excited state; M, the reduced mass of the molecule; and $d\mu/du$ the effective dynamic charge e^* of the dipole. With these approximations, the differential inelastic scattering cross section for a single loss (i.e. $\nu = 0 \to 1$) is

$$\frac{d^2\sigma}{d\Omega dE_1} = \frac{1}{A^2 S_{eff}} \left(\frac{E_1}{E_0}\right)^{\frac{1}{2}} \frac{m}{M} \frac{e^2}{E\hbar\omega_0} \left(\frac{d\mu}{du}\right)^2 |f(\vec{k}_0,\vec{k}_1)|^2 \; \delta(E_0 - E_1 - \hbar\omega_{vib}) \quad (16a)$$

with $E_0 = E_1 + \hbar\omega_{vib}$, and $f(\vec{k}_0,\vec{k}_1)$ is a dimensionless scattering amplitude

$$f(\vec{k}_0,\vec{k}_s) = \frac{k_0}{4\pi} < \psi_{k_s}^- | \frac{\vec{\mu}_0 \cdot \vec{r}}{r^3} | \psi_{k_0}^+ > \quad (16b)$$

which is readily evaluated once $A(\vec{k})$, $\delta(\vec{k})$ and e^* are chosen [1]. $S_{eff} = S \cos \theta$, where S is the surface area per adsorbed molecule and θ is the electron impact polar angle. The factor A^2 normalizes the distribution (16a) per electron elastically reflected in the specular direction.

3.3.1 Modes Perpendicular to the Surface

Assuming perfect reflectivity from an infinite surface potential barrier (i.e. $A^2 = 1$ and $\delta = \pi/2$), LENAC et al. [7] produced numerical results for the vibrational scattering amplitudes corresponding to the CO stretching mode, $\hbar\omega_{CO} \sim 0.2$ eV, for the molecular dipole adsorbed perpendicularly on the surface. Their results show that the largest contributions arise from processes 1 and 2 (Fig.4). Whereas inelastic scattering from the free dipole (Fig.3) is peaked about the forward scattering direction, in the case of the adsorbed dipole, the inelastic intensity peaks strongly about the specularly reflected beam. An interesting feature is that, in general, two lobes of intensity occur in the incident plane, one on either side of the specular beam (fig.5a). The relative intensity of these two lobes depends sensitively

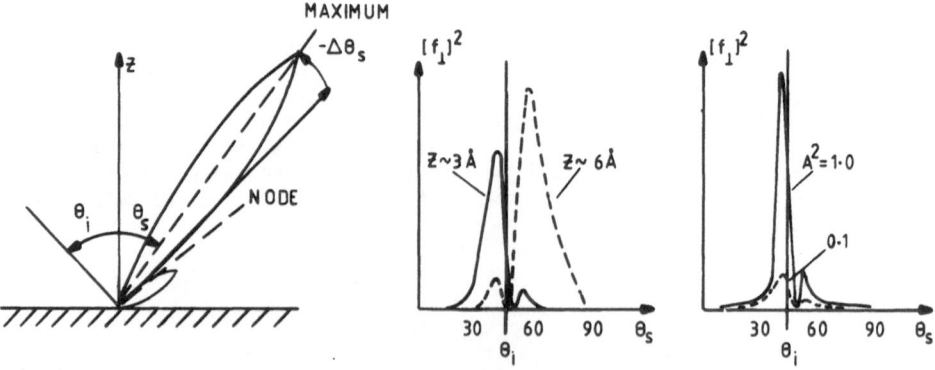

Fig. 5. Angle dependence of the dipole scattering intensity appropriate to the C-O stretching vibration for the molecule oriented perpendicular to a semi-infinite metal surface [7] (a) polar angle plot in the plane of incidence; $\theta_i = 45°$. (b) differential cross section as function of scattering angle θ_s for two positions above the image plane: $z = 2k_0^{-1}$ (full line); $z = 4k_0^{-1}$ (dashed line): ($k_0 = 0.6$ Å^{-1} for $E_0 = 1.32$ eV, $\hbar\omega_1 = 0.25$ eV) (c) variation of the differential cross section with reflectivity: $A^2 = 1.0$ (full line); $A^2 = 0.1$ (dashed line).

on the position fo the dipole above the surface (Fig.5b) and on the electron reflectivity (Fig.5c). When, as is usually the case for adsorbates, the dipole is very close to the surface, the lobe closest to the surface normal direction is the larger. The angle of displacement of the lobe maximum from the specular direction $-\Delta\theta_s$ is a function of $\hbar\omega_1/2E_0$, where $\hbar\omega_1$ is the single loss energy and E_0 the incident impact energy [1]. This lobed intensity distribution is a result of our choice of specularly reflected plane waves (14) which have a node at the surface due to the position of the dipole suppressing the specularly reflected peak [7].

The essential correctness of these theoretical results is shown by the experimental work of ANDERSSON et al. [18]. Fig.6a shows a plot of their electron energy loss results for the variation in intensity of the fundamental C-O stretching mode in a c(2x2) CO monolayer on Cu(100) as a function of collection angle scanning away from the specular beam direction $\pm\ \Delta\theta_s$. The experimental data (open circles) confirm that the inelastic cross section is sharply peaked in a direction off specular in agreement with the dipole model calculation (dashed curve). The calculated curves for the differential inelastic scattering cross section (16a) were evaluated by integrating over a collection aperture solid angle corresponding to a cone of half angle 3° and a surface density of non-interacting oscillating dipoles $N_j \simeq 1.19 \times 10^{18}$ molecules per cm^2.

When (16) is integrated over the full half-space solid angle Ω, the loss signal increases as $E_0^{-3/2}$ with decreasing incident energy. However, for the small collection angles appropriate to the analyzers in current use, the shift of the loss intensity away from the specular direction at low excitation energies tends to dominate the $E_0^{-3/2}$ behaviour such that the ratio of the loss intensity to the elastic intensity decreases as $E_0 \to 0$. This is seen in the results for the impact energy dependence of the differential cross section, Fig.6b.

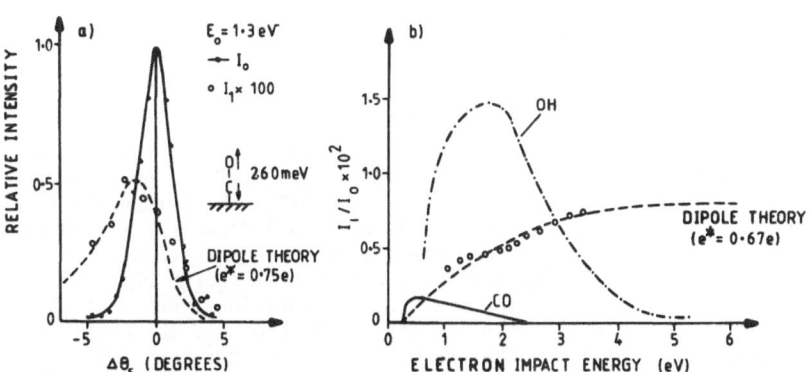

Fig.6 (a) Angular dependence of elastic beam I_0 and inelastic loss I_1 (260 meV) intensities for c(2x2) CO on Cu(100) [18]
(b) Impact energy dependence of the relative loss I_1/I_0, for c(2x2) CO on Cu(100) [18]
Dashed curves (dipole theory); full curve (negative ion resonance for adsorbed CO [19]); chain curve (negative ion resonance for OH [21]).

The experimental points are again seen to obey closely the behaviour expected for dipole scattering (dashed curve). However, there is some (small) deviation from the theoretical curve below $E_0 \approx 2$ eV. PERSSON [19] has remarked on this feature and suggested that, amongst the number of physical processes neglected in the dipole theory (16), the observed deviation can be explained in terms of short range scattering arising from the formation of a temporary negative ion resonance on the CO equivalent to the inelastic scattering processes 3 or 4, Fig.4. PERSSON [19] has estimated the contribution to the ratio I_1/I_0 to be that shown in Fig.6b which is of the right order of magnitude to explain the deviation of the experimental points over and above the dipole curve. The Xα-multiple scattering calculation which PERSSON employed assumes an oriented CO molecule, the orbitals of which are relatively unperturbed by the bonding to the surface [4]. However this is known not to be the case. CO chemisorption on copper involves the donation of charge to the metal from the 5σ orbital with back donation into the molecules $2\pi^*$ antibonding orbital [20]. As a consequence, one expects this resonant antibonding orbital to become partially filled and modified. The question that remains is the extent to which this back-bonding from the metal will prevent temporary negative ion formation occurring at all. PERSSON's analysis [19] would suggest that there may be some residual effect remaining. However, since the incident electron is accelerated by its own image charge to the metal surface [1] one does not expect the resonance energy to be the same as the gas phase value $E_0 \sim 1.7$ eV [4]. The "fit" with the experimental deviation from the dipole curve (Fig.6b) would suggest a value closer to $E_0 \sim 0.5$ eV for the adsorbed molecule.

Perhaps a more clear-cut case of temporary negative ion scattering involving an adsorbate molecule is the case of the O-H vibrational loss observed by ANDERSSON and DAVENPORT [21] for OH radicals on NiO(111). In this case, the deviation from dipole behaviour is strongly marked, as shown in Fig.6b. The oxidized nickel surface forms a thin "insulating" layer (~ 6 Å thick) and it would seem likely that the antibonding levels of the OH are located at energies corresponding to the energy band gap of the insulator. This being the case, the resonant state has a long lifetime since the appropriate molecular orbitals are unable to hybridize with those of the substrate.

3.3.2 Surface Trapping in Bound State Surface Resonances

One of the assumptions of the above dipole theory (16) is that the electron reflectivity and phase shift do not differ appreciably before and after the electron dipole interaction. That is, if $A(E_0) \simeq A(E_0 - \hbar\omega_1)$ then the ratio of the inelastic to the elastic signal I_1/I_0 will be independent of the substrate reflectivity A. This is a good approximation for copper and nickel surfaces, the variation of $A^2(\vec{k}_0)$ and $\delta(\vec{k}_0)$ with impact energy being a rather smooth function, particularly on the scale of CO vibrational energies, $\hbar\omega_1 = 0.25$ eV [7]. I_1/I_0 is then a slowly varying function of E_0, the magnitude of which is determined solely by the dynamic effective charge e* (Fig.6b). However, Fig.7 shows an example where this is certainly not the case, viz. inelastic scattering from the symmetric stretch hydrogen mode normal to the surface of a W(100)p(1x1)H layer (corresponding to saturation coverage)[22].

In Fig.7a we see that a condition arises in which the reflectivity from the W(100) surface fluctuates rapidly for impact energies in the range 3 to 6 eV [23]. When this measured reflectivity is incorporated into the scattering from a dipole potential (16), I_1/I_0 deviates strongly from the behaviour predicted by dipole theory (dashed curve, Fig.7b). This behaviour

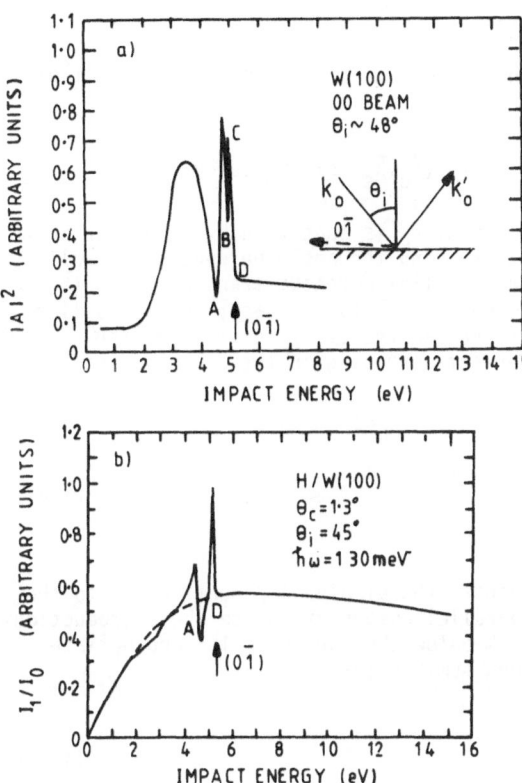

<u>Fig.7</u> Illustrating the effect of a rapid change in electron reflectivity,
$A(E_0) \neq A(E_0 - \hbar\omega_1)$, on the impact energy dependent scattering
intensity of the symmetric stretch hydrogen mode on W(100).
(a) Measured reflectivity, $|A|^2$ [23]
(b) Calculated dipole scattering cross section with (full line) and
without (dashed) incorporating $|A|^2$; aperture solid angle θ_c.
Sharp minima in the reflectivity A, B, C D, associated with the
grazing emergence of the (01) diffraction beam, produce sharp
maxima in the normalized loss curve, I_1/I_0 [22].

arises from the grazing emergence condition for new diffraction beams; at
the energy indicated by the arrow (Fig.7b) and for the geometry used in
this particular experiment ($\theta_i = 48^0$, $\Delta\theta_s = 0^0$), the $0\bar{1}$ first order
diffraction beam just emerges along the surface (inset, Fig.7a). If, as is
the case here, there is an energy band gap in the metal's distribution of
electronic states [24], the electron may be trapped at the surface in a
state characteristic of the one-dimensional "image" potential well. These
so-called "Rydberg states" are quantized in momentum normal to the surface
and possess relatively long lifetimes (note, the widths of the fine structure
fluctuations, Fig.7b). Tuning the electron spectrometer into one of these
states produces surface electron waves which propagate over the surface and
enhance considerably the vibrational excitation cross sections in the process
[22]. These *surface resonances* (A,B,C,D etc., Fig.7a) are the two-dimensional surface analogues of the negative ion resonances in the gas phase [6],
the only difference being that the former reflects the symmetry of the crystal
surface whereas the latter possesses molecular symmetry.

3.3.3 Modes Parallel to the Surface

The scattering amplitude (16b) can be decomposed into components parallel and normal to the surface [7]

$$f(\vec{k}_0,\vec{k}_s) = f_\perp(\vec{k}_0,\vec{k}_s)\cos\theta_0 + f_{//}(\vec{k}_0,\vec{k}_s)\sin\theta_0\cos\beta \qquad (17)$$

where the direction of the dipole moment $\vec{\mu}_0$ is specified by the polar angle θ_0 and azimuthal angle β, this latter being the angle between the parallel component of the dipole $\vec{\mu}_{//}$ and the parallel momentum transfer $q_{//} = (\vec{k}_0-\vec{k}_s)_{//}$. However, when the molecule is adsorbed on a metal surface, the conduction electrons in the metal act so as to screen the electric field of the vibrating dipole. If we assume perfect screening in the close vicinity of metal surfaces [1]

$$\vec{\mu}_{//} = -\vec{\mu}_{//}(image)$$
$$\vec{\mu}_\perp = \vec{\mu}_\perp(image) \qquad (18a)$$

i.e. the screening effectively enhances the dipole component normal to the metal surface and suppresses the parallel component due to the introduction of "image" dipoles at a distance $-z_0$ below the surface. The appropriate image and real scattering amplitudes sum, so that

$$f_\perp^{total} = f_\perp(z_0) + f_\perp(-z_0)$$
$$f_{//}^{total} = f_{//}(z_0) - f_{//}(-z_0) \qquad (18b)$$

and in the limit $z_0 \to 0$,

$$f_\perp^{total} \to 2f_\perp(0) ; \qquad f_{//}^{total} \to 0 \qquad (18c)$$

This result corresponds to the well-known selection rule in infrared spectroscopy of adsorbed molecules on metallic surfaces which states that only those modes that give a net dipole change perpendicular to the surface can absorb IR radiation [25]. In this case, the photon wavelengths are so long ($z_0 \ll \lambda_{IR}$) that the adsorbed dipole together with its image effectively "see" the same field. However, in the case of incident electrons with one electron volt impact energy, the electron wavelength is only about 2 Å so that $z_0 \approx \lambda_e$ and strong interference effects arise between the scattering processes (Fig.4) and between scattering from the normal and parallel components of the dipole. For $z_0 = 0$, the contribution from the normal dipole is four times larger in intensity when its image is included while the parallel contributions vanish (18c). However, as the dipole is removed from the surface ($z_0 \neq 0$), the calculated total cross sections [7] fluctuate appreciably. As shown in Fig.8a-c, for various orientations of the point dipole, the inelastic cross sections are sensitive to the assumed choice of z_0 (though the variations may be somewhat exaggerated because of the approximation of scattered plane waves (14) in this particular case). In order to illustrate the possible importance of the effect, the positions of the centre of gravity of CO adsorbed on nickel surfaces in top (T), bridge (B) and

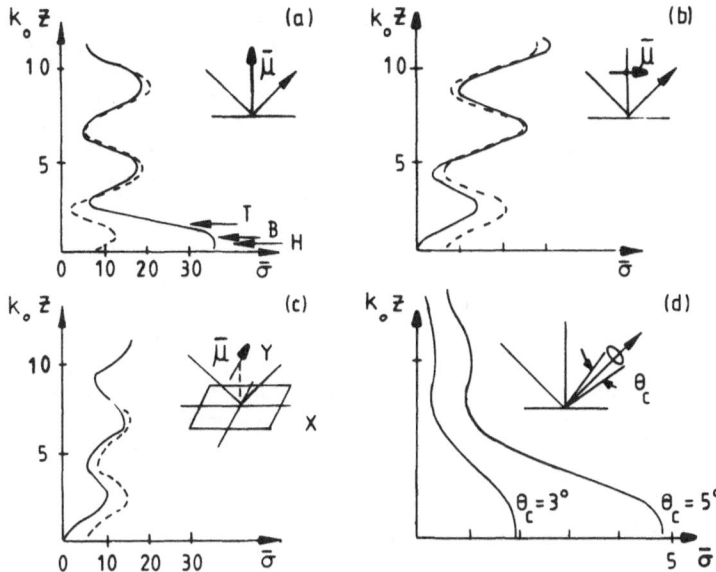

<u>Fig.8</u> Dependence of the total cross section $\bar{\sigma}$ upon the normalized distance ($k_0 z$) of the dipole from the surface for the three principal orientations:
(a) perpendicular to the surface; (b) parallel, in the plane of incidence; (c) parallel, but normal to the plane of incidence; (d) as in (a) for two values of the aperture acceptance angle, θ_c. Broken curves are without the image potential. T = top (2.5 Å); B = bridge (1.2 Å); H = hollow (1.05 Å) for CO on Ni(100) [7].

hollow (H) sites are indicated. We see that while scattering from perpendicular modes will dominate for distances typical of *atomic* adsorbates, parallel dipole cross sections do increase quite markedly at larger distances, $z_0 \approx 2.5$ Å. This will be important in scattering from large molecular adsorbates. Figure 8d illustrates the fact that this variation is not an artifact of large angle scattering; the differential scattering cross section integrated over a spectrometer slit solid angle of $2\theta_c$ corresponging to 6° and 10° respectively shows that, apart from the expected overall decrease, the z_0 dependence remains.

3.4 Dipole Theory of One-Phonon Excitation

Of course, it is not possible in an actual experiment to observe the vibrational modes of a single adsorbate molecule; the spectra are usually recorded as a function of coverage. An alternative approach to the above surface dipole theory of scattering from a single adsorbate is to use a form of the time dependent Born approximation, as introduced by Evans and Mills [13], to describe inelastic electron scattering from the long range dipole field fluctuations of surface optical phonon modes. This approach may be extended to adsorbate modes, i.e. the adsorbed layer is treated as an extended array of (non-interacting) point charges located a small distance above the image plane, z_0 [8]. The electrostatic interaction is assumed to occur directly between the incident electron and the long range fields of

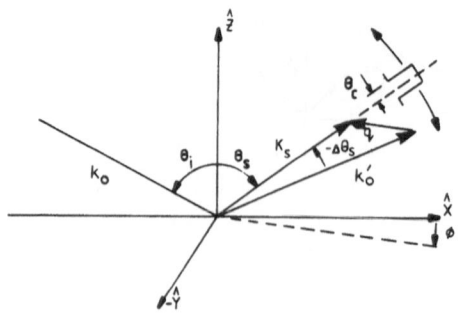

Fig.9 Scattering geometry for surface phonon excitation, wavevector $\vec{q}_\| = (\vec{k}_0 - \vec{k}_s)_\|$. Scattering direction $-\Delta\theta_s$ refers to measurement from the specular beam direction k_0' towards the surface normal.

the adsorbate's phonon modes, and indirectly via the image charge polarization of the metal surface. Again using first-order, time-dependent perturbation theory, the differential cross section for a given vibrational mode per adsorbate takes the form [8]

$$
\frac{d\sigma}{d\Omega dE_1} = \left(\frac{1}{2\pi}\right)^2 \left|\frac{\vec{k}_s}{\vec{k}_0}\right| \cos\theta_s \left\{ \left| \vec{u}_\perp f_\perp(|\vec{q}_\|| , k_\perp^0, k_\perp^S) \right. \right.
$$
$$
\left. \left. + \; \vec{u}_\| \cdot \vec{q}_\| \; z_0 f_\|(|\vec{q}_\||, k_\perp^0, k_\perp^S) \right|^2 \right\} \; \delta(E_0 - E_1 - \hbar\omega_1)
$$

(19)

where $\vec{q}_\| = (\vec{k}_0 - \vec{k}_s)_\|$ represents the change in the incident electron's wavevector parallel to the surface arising from the excitation of a phonon mode. The scattering geometry is shown in Fig.9. The wavefunctions of the incident $\vec{k}_0$ and scattered $\vec{k}_s$ electrons take the plane-wave form (14) since the scattering is assumed to occur outside the metal surface. $\vec{u}$ is the oscillator displacement amplitude for modes perpendicular ($\vec{u}_\perp$) and parallel ($\vec{u}_\|$) to the surface. The differential scattering probability is obtained by multiplying (19) by the number of adsorbates per unit area N and summing over all degenerate vibrational modes L corresponding to the adsorbate lattice symmetry

$$
\frac{dP}{d\Omega} = N\sum_{i=1}^{L} \frac{d\sigma^i}{d\Omega}
$$

(20)

where $d\sigma^i/d\Omega$ is the differential cross section for the ith mode. The numerical results of ROSE and WILKINS [8], showing the angular dependence of the inelastic electron scattering from a *monatomic* adsorbate vibrating either parallel or perpendicular to the surface, are summarized in Fig.10.

It can be seen that the scattering is much stronger for the perpendicular modes (Fig.10a), as is expected by dipole theory. The contour plot of the scattered intensity shows that there is an intense lobe of intensity with its maximum (marked M) followed by a node (marked N) in the plane of incidence, similar to that predicted for single dipole scattering earlier (Fig.5). In real space, these maxima form a distorted ring-like structure around the minimum close to the specular beam direction. The node N is seen to be a saddle point in the azimuthal plane. The scattering from vibrations parallel to the surface (Fig.10b and c) is roughly $(q_\| z_0)^2$ weaker in intensity [8]. The angular distribution from a single adsorbate depends

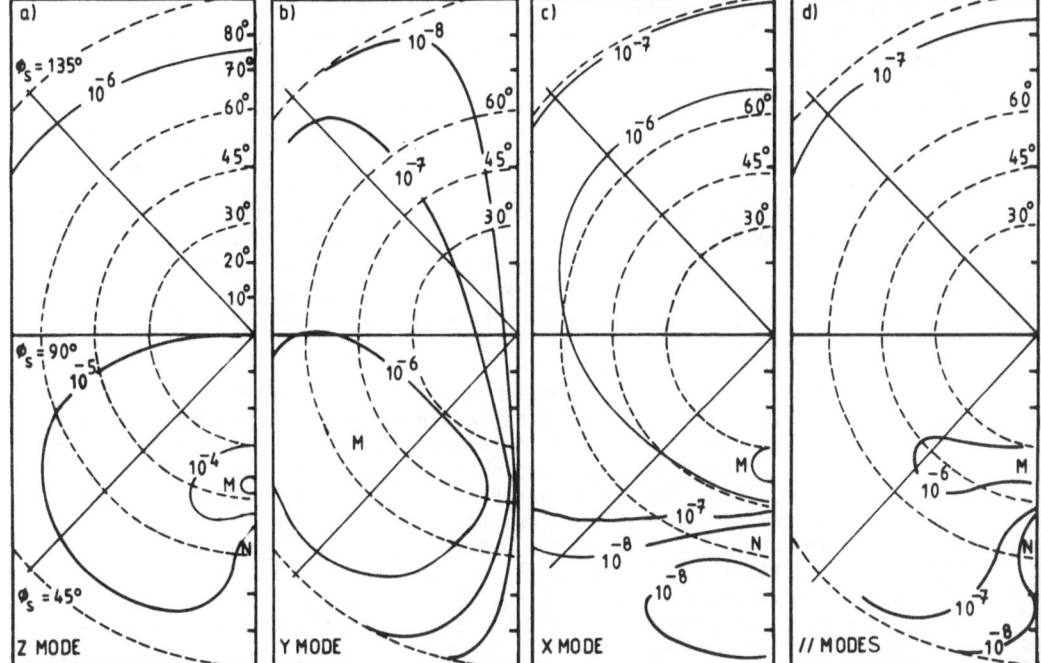

Fig.10 Loss intensity contour plots for dipole scattering from the vibration
modes of a monatomic adsorbate [8]
(a) perpendicular displacements
(b) parallel displacements along y-axis (normal to plane of incidence)
(c) parallel displacements along x-axis (plane of incidence)
(d) parallel modes of square lattice array (averaged over all parallel
    directions)
Polar angle θ_S is along the ordinate axis; azimuthal radial angles
are shown. M = lobe maximum; N = node
$\theta_i = 45^0$; $E_0 = 1.5$ eV; $\hbar\omega_1 = 50$ meV.

Fig.11 Lattice modes of H chemisorbed on W(100) at sat-
uration coverage. Dis-
placement of H atoms
lateral to the surface
is indicated by arrows;
vertical motion towards
vacuum (+) and into the
surface (-).

strongly on the orientation of the adsorbate's vibrational axis; Fig.10b shows a contour plot for displacement along the y-axis, and Fig.10c is that along the x-axis (z axis assumed normal to the surface). If one sums the scattering intensities over all the adsorbates species for all degenerate modes, the azimuthal scattering intensity tends to average out. Fig.10d shows the scattering intensity calculated for a monatomic adsorbate such as hydrogen in the square lattice array, Fig.11. The important difference is that the scattering is much more diffuse from the parallel modes. Also, the angular distributions present a complicated pattern around the specular direction, although the maxima M are much less pronounced than is the case for the perpendicular modes. For a given incident energy E_0, the overall intensity of the perpendicular modes scales as $1/(\hbar\omega_\perp)^3$, i.e. the vibrational modes with small energy losses are much emphasized. For the parallel modes, the intensity scales only as $1/(\hbar\omega_{//})$ [8].

In Fig.12, we show that the signal strength of both the perpendicular and parallel modes can be increased by inclining the incident beam towards grazing incidence [1]. The dashed lines represent the scattering intensity integrated over a small slit aperture solid angle centred on the lobe intensity maximum. The solid lines represent the specular direction. It can be seen that for incidence angles less than $\theta_i \approx 45^0$, the signal may be substantially improved by moving the detector to the position of maximum intensity, the position of which will depend on the substrate metal reflectivity and the ratio, $\Delta E_{loss}/E_0$. Also, the scattering efficiency goes to zero for very large angles [1]. This suggests an optimum incidence angle for maximum scattering intensity $\theta_i \approx 80^0$ for observing *all* modes.

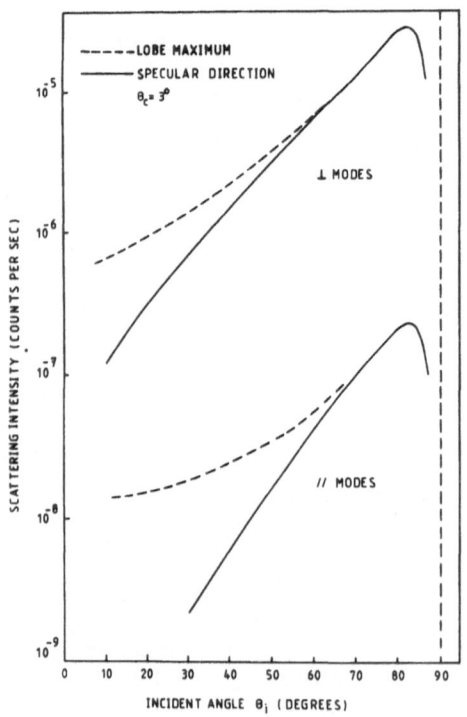

Fig.12 Calculated signal strengths for modes vibrating perpendicular ($\perp$) and parallel ($//$) to the surface for scattering into a circular aperture subtending a half angle of 3^0 [8].

38

For vibrations parallel to the surface, the electrons are scattered primarily by the oscillating electric quadrupole interaction, the relative short range of which (compared with dipole scattering) gives rise to larger momentum transfers and the diffuse nature of the scattering. The potential due to the charge distribution at each lattice site $\vec{X}_{//}$ together with its image may be written [9]

$$\Phi(r) = 2\sum_{\vec{X}_{//}} \left| \frac{\mu_z r_z}{|\vec{r} - \vec{X}_{//}|^3} + \frac{Q_{xz}r_x r_z + Q_{xy}r_y r_z}{|\vec{r} - \vec{X}_{//}|^5} \right| \tag{21}$$

where $\Phi(r)$ is the potential far from the surface for a periodic array of dipoles (normal component μ_z) and quadrupoles (Q_{xz} and Q_{xy} being the corresponding elements of the quadrupole moment tensor). Eq. (21) may be expanded to first order in displacements, which are assigned the usual phase factor $\exp(i\vec{q}_{//} \cdot \vec{X}_{//})$ for a surface (phonon) wave of wavevector $\vec{q}_{//}$. THOMAS and WEINBERG [9] applied this approach to derive a general analytical expression for the differential scattering probability, including separate dipolar and quadrupolar contributions. Their results for the hydrogenic lattice modes of H adsorbed on W(100) at saturation coverage (represented in Fig.11) are shown in Fig.13 as a function of scattering angle about the specular beam, $\Delta\theta_s$. As observed by HO et al. [2], the asymmetric stretch (ν_{as} or x-mode) and wagging (ν_w or y-mode) vibrations parallel to the surface produce a relatively isotropic distribution, whereas the perpendicular vibration (ν_s or z-mode) peaks strongly about the specular direction and dominates the intensity distribution overall due to the stronger dipole term

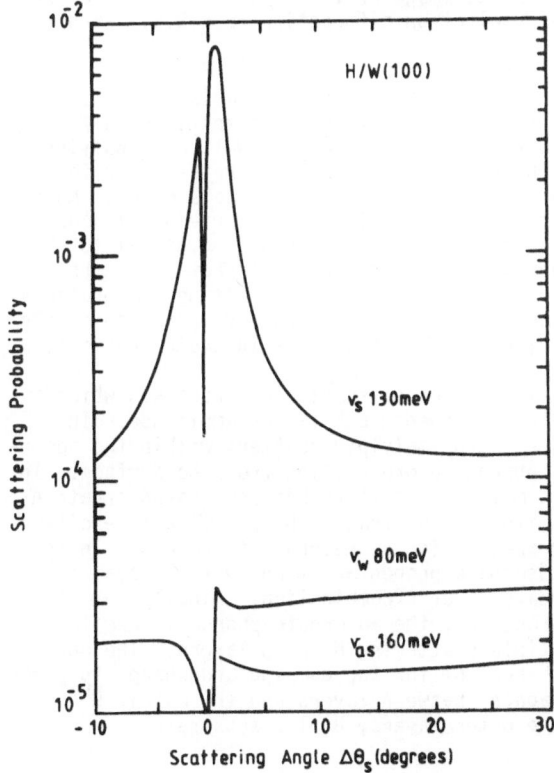

Fig.13

Scattering probability as a function of scattering angle $\Delta\theta_s$ relative to the specular beam direction for the symmetric (ν_s), asymmetric (ν_{as}) and wagging (ν_w) modes of p(1x1)H on W(100) [9].

$\vec{\mu} \cdot \vec{r}/r^3$ (21). Also, *all* modes show a node close to the specular direction in agreement with the dipole lattice thoery [8].

3.5 Multiple Scattering Theory of One-Phonon Excitation

One important ingredient in the overall theory has been neglected to date - multiple scattering of the incident electrons within the outermost layers of the adsorbate-substrate lattice. As the electron approaches very close to the surface, it is accelerated by its dipole image charge. One consequence of this is that we define a "cutoff distance" at which the long range dipole screening mechanism breaks down; the incident electron effectively breaks through the metal surface's electronic screening charge to impact directly with the thermal motion of the adsorbate-substrate surface atoms. The short range (impact) scattering potential of the ion cores produces diffuse scattering over a much wider angular range than is the case in dipole scattering, although the latter acts over a much longer time scale, $\tau_{LR} \simeq 100 \; \omega_{SR}$ [19]. Another consequence is that the incident electron is able to penetrate into the surface region and undergoes multiple scattering between the individual adsorbate and substrate atoms. Here there are large phase differences between scattered (Bloch) waves which originate from different regions of the structure, and the interaction of the incident electron with the substrate can no longer be viewed simply as a single reflection event from an infinite potential barrier in which the scattering occurs from a point dipole potential plus its image [7]. The electron scattering now relates formally to low-energy-electron-diffraction (L.E.E.D.) theory [26].

LI, TONG and MILLS [10] have used this approach to formulate a wide-angle electron impact thoery which utilizes muffin tin potentials to describe the multiple scattering from the ion cores, as well as embodying an explicit description of the interaction with the adsorbate-substrate vibrating system via one-phonon excitation of the adsorbate's modes. Numerical calculations [27] show that the large-angle spectra provide all the normal modes, while near-specular data allow modes polarized normal to the surface to be selected. An important result is that diffraction of the electron by the surface layers either before or after emitting a vibrational quantum produces pronounced structure in the impact energy variation of the loss intensity. In what follows, we outline the steps in their argument.

Figure 14 summarizes four basic classes of scattering processes which enter the final expression for the one-phonon excitation amplitude. In the first process, the incoming electron multiply scatters within the adsorbate layer A only and excites a phonon before exiting from the surface. In the second process, the electron multiply scatters between the adsorbate A and outermost layer B combined with the substrate S before finally exciting an adsorbate phonon on exit. Process 3 is the reverse of process 2 in that the electron first excites an adsorbate phonon and then multiply scatters off the layer B/substrate/adsorbate-layer A combination. Finally, in the 4th combination of scattering processes, the adsorbate phonon is excited sometime during the overall multiple scattering between layers. The one-electron propagators $\underline{P}^{AB}$ and $\underline{P}^{BA}$ account for the damping and change in phase of the electron wave as it propagates between layers and the matrix $\underline{R}^{BS}$ describes its reflection from the outer layers, B plus substrate.

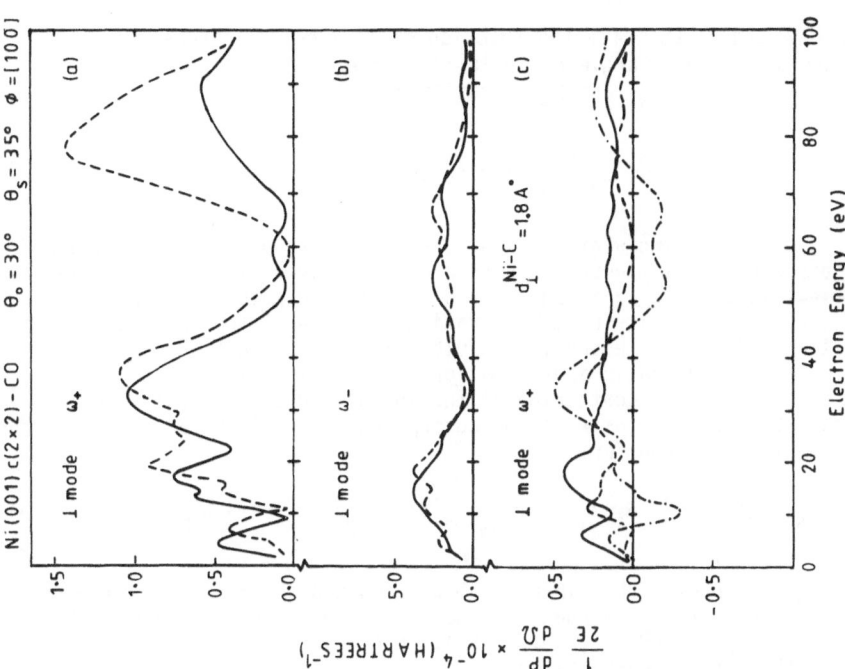

Ni(001) c(2×2)-CO $\theta_o = 30°$ $\theta_s = 35°$ $\phi = [100]$

(a) ⊥ mode ω_+

(b) ⊥ mode ω_-

(c) ⊥ mode ω_+ $d_\perp^{Ni-C} = 1.8 \text{Å}$

Electron Energy (eV)

$\frac{1}{2E} \frac{dP}{d\Omega} \times 10^{-4}$ (HARTREES^{-1})

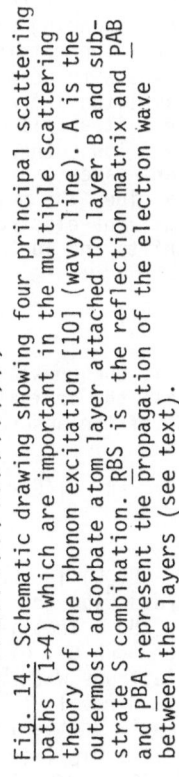

LAYER A
LAYER B
SUBSTRATE

Fig. 14. Schematic drawing showing four principal scattering paths (1→4) which are important in the multiple scattering theory of one phonon excitation [10] (wavy line). A is the outermost adsorbate atom layer attached to layer B and substrate S combination. $\underline{R}^{BS}$ is the reflection matrix and $\underline{P}^{AB}$ and $\underline{P}^{BA}$ represent the propagation of the electron wave between the layers (see text).

Fig. 15. Imapct energy dependence of the differential loss probability for the perpendicular stretching modes of c(2x2)CO on Ni(100); C-O stretch ($\omega+$), Ni-C stretch ($\omega-$). $d_\perp$(Ni-C) = 1.8 Å (solid curves) and 1.06 Å (dashed curves). Separate contributions due to scattering from the O atoms (solid curve), C atoms (dashed curve) and interference effects (chain curve) are shown in the lower pannel (c) [27].

41

LI, TONG and MILLS write the fraction of incident electrons which are scattered onto a solid angle $d\Omega$ following the excitation of a vibrational mode with wavevector $\vec{q}_{/\!/}$ and polarization j, as

$$\frac{dP(\vec{k}_0,\vec{k}_s)}{d\Omega} + \frac{mE_0}{2\pi^2\hbar^2}\frac{\cos^2\theta_s}{\cos\theta_0} \ S|M\ (\vec{k}_0,\vec{k}_s)\ \vec{q}_{/\!/,j}| \tag{22}$$

with θ_0 and θ_s, the incident and scattered angles respectively (Fig.9) and S the sample normalization area. For an ordered layer of adsorbate molecules, the surface is considered as an array of muffin tin potentials, each diplaced from its equilibrium site by the vibrational motion of the respective atoms. The initial and final electronic and vibrational wavefunctions are again separable in the adiabatic approximation (cf. (2) and (15)) such that the matrix element for the scattering of the incident electron from $|\vec{k}_s>$ to $|\vec{k}_f>$ with the creation ($\sigma_i = +1$) or absorption ($\sigma_i = -1$) of an array of phonons ($\sigma_i q_{/\!/}^{(i)}$) takes the form

$$M(\vec{k}_0,\vec{k}_s; \ \{\sigma_i q_{/\!/}^{(i)}; \ \alpha_i\}) \ = \ <\{n_{\vec{q}_{/\!/}}\alpha\}_s|f(\vec{k}_s,\vec{k}_0;\{\vec{R}\}\ | \ \{n_{\vec{q}_{/\!/}}\alpha\}_0> \tag{23}$$

where $|\{n_{\vec{q}_{/\!/}}\alpha\}>$ denotes the vibrational wavefunctions for modes with quantum number α, dependent on the instantaneous positions of all the nuclei in the system $\{\vec{R}\} = (\vec{R}_1 \ .... \ \vec{R}_M)$. For an adsorption site of high symmetry, the normal modes may be decomposed into motion perpendicular to the surface, $q_{/\!/}^{(i)}$ referring to the ith Cartesian component of a displacement $R^{(i)}$ corresponding to a mode with wavevector $q_{/\!/}^{(i)}$ propagating in the surface layer. The matrix element (23) may then be expressed in terms of the derivatives with respect to nuclear displacements $\delta f(\vec{k}_s,\vec{k}_0)/\delta R$ of the scattering amplitudes $f(\vec{k}_s,\vec{k}_0)$ summed over the adsorbate atoms. The scattering process itself is described in terms of a formalism developed within the context of low energy electron diffraction (LEED) theory [26]. $|k_0>$ and $|k_s>$ are plane waves that describe the incoming and outgoing electron, and a T matrix describes the multiple scattering within the surface layers combined with the usual one-electron Green's funciton $G(\vec{r},\vec{r}')$ corrected for inner-potential shifts and inelastic damping effects

$$f(\vec{k}_s,\vec{k}_0; \ \{R\}) \ = \ <\vec{k}_s \ |GT| \ \vec{k}_0> \tag{24}$$

If $V(\vec{r},\{\vec{R}\})$ is the crystal potential with the atoms displaced from their lattice sites, the derivatives of the scattering amplitudes with respect to displacement of the adsorbate atoms takes the form

$$\left[\frac{\delta f}{\delta R^{(i)}}\right] \ = \ <\vec{k}_s \ |(G + GT_0G)\left[\frac{\delta V}{\delta R^i}\right] \ (1 + GT_0)|\vec{k}_0> \tag{25}$$

The term $(G + GT_0G)$ is the same one particle propagator g_{PE} that is employed to describe the outgoing electron wave in angle-resolved photoemission theory [28]. That is, if an electron is emitted from some point in the lattice, then $<\vec{k}_s|G + GT_0G)|$ describes its propagation to a detector oriented to accept electrons with wavevector $\vec{k}_s \equiv \vec{k}_f$, of the final state. The electron may propagate directly ($<\vec{k}_s|G>$) or it may arrive at the detector only after multiple elastic scattering from the lattice ($<k_s|GT_0G$). Similarly, the combination $(1 + GT_0)|\vec{k}_0>$ is precisely the initial state wavefunction that enters LEED theory [26] to describe the incoming electron, which propagates to a point in the surface lattice (engaging in multiple scattering along the way) where it interacts with a displaced adsorbate atom through the term $[\delta V(\{\vec{R}\})/\delta R]$. If we now choose to define an EELS wavefunction

$$|\psi_{EELS}\rangle = g_{PE} \left[\frac{\delta V}{\delta R^i}\right] |\psi_{LEED}\rangle \tag{26}$$

then (25) becomes

$$\left(\frac{\delta f}{\delta R^i}\right) = \langle\vec{k}_s|\psi_{EELS}^{(i)}\rangle \tag{27}$$

The importance of this expression lies in the fact that once we are able to develop methods for calculating the potential of the derivative function $(\delta V/\delta R^i)$, the computational technology recently developed for LEED and angle-resolved photoemission intensity theory may be readily utilized.

TONG, LI and MILLS [27] have determined the quantity $(\delta V/\delta R^i)$ for an AB overlayer, with the substrate treated as rigid, corresponding to a c(2x2) layer of CO adsorbed on Ni(100) (re: Fig.6). For the top adsorbate atom, A, $(\delta V/\delta R_A^i)$ was determined by shifting the appropriate muffin tin potential centered at $\vec{R}_A$. The T matrix depends parametrically on the instantaneous positions of the ion cores in the adiabatic approximation. The incident electron scatters from the associated array of potentials, with the ith potential displaced slightly from the equilibrium position, $\vec{R}_i \neq R_i^{(0)}$. The resulting T matrix directs the electron away from the Bragg diffraction directions by virtue of the disorder associated with the displacements. For each normal mode of the molecule, the matrix element (23) is a coherent super-position of scattering amplitudes from atoms A and B, and since the cross section (22) is proportional to M^2, there are interference terms in addition to the sum of the squares of the individual amplitudes. In general, there-fore, for fixed incident and scattered electron directions, we expect distinctly different energy dependences of the loss cross sections for scattering from the different modes of the adsorbed molecule AB.

This is illustrated in the calculated energy dependence of the differen-tial loss probability curves for the normal stretching modes of c(2x2)CO on Ni(100), Fig.15. Here the variation of the one-phonon loss cross sections for the C-O stretching mode with frequency (ω_+) and the C-Ni with frequency (ω_-) for vibrations normal to the surface are shown as a function of electron impact energy for a scattering angle $\Delta\theta_s \simeq 5^0$ off specular. For a diatomic molecule oriented vertically against a rigid lattice, the atomic masses vibrate out of phase in the case of the high frequency ω_+ mode, and in phase in the case of the low frequency ω_- mode. The form of the scattering amplitude coefficients (25) will be different which will produce different interference cross terms in the overall scattering probabilities (22). In Fig.15, each mode has been calculated for two bonding sites - top sites (T) and hollow sites (H). The electron impact cross section for the ω_+ mode is highly sensitive to the Ni-C spacing of the two bonding sites illustrating the importance of diffraction of the incident and scattered beams between the overlayer and substrate atoms. There is less structure in the scattering from the C-Ni ω_- mode, though again the magnitude and energy variation of the loss intensity is site sensitive.

Each mode is described by a loss cross section that is a distinctly different synthesis of the scattering amplitude derivatives (27), and each may in general be expected to have its own bonding site dependence. Thus, the impact energy dependence of EELS data at off specular scattering angles provides a rich source of surface structural information.

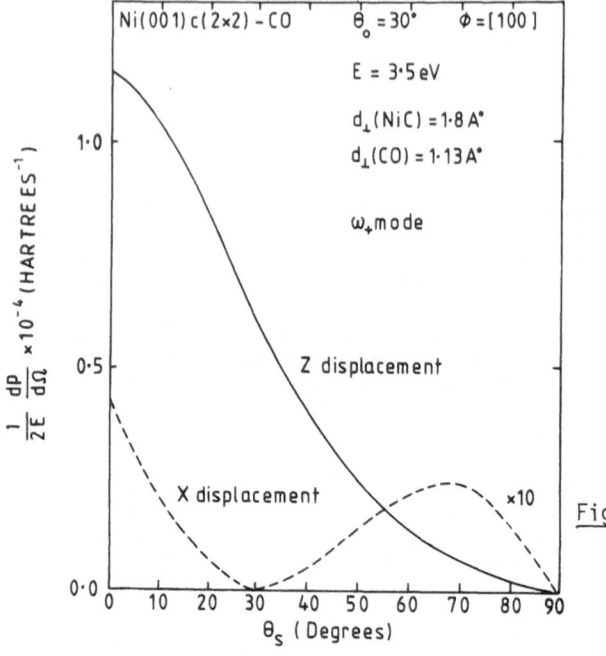

Fig.16 Scattering angle dependence of the differential loss probability for displacements perpendicular (z) and parallel (x) to the surface [27].

Fig.16 shows the angular variation of the loss cross sections for in-plane scattering ($\vec{k}_0, \vec{k}_s$ and the surface normal direction $\hat{z}$ in the same plane) from the C-O stretching (ω_+) modes perpendicular (z displacement) and the parallel (x displacement) to the surface. The incident angle, $\theta_0 = 30°$. For impact scattering, the electrons are broadly distributed in angle but note the zero in intensity at the specular direction for scattering from the parallel mode. SEBASTIAN [29] has recently commented on this feature and has shown that it is a consequence of reflection symmetry through the scattering plane for adsorbates occupying high point-group symmetry sites. Parallel modes polarized normal to the scattering plane scatters with zero intensity whenever $\vec{k}_s$ is swept through the scattering plane, not only along the specular direction, as in Fig.16. This remains true even in the case of surface bound state resonance scattering [22]. These nodes in the angular distribution of the scattered intensity can be used to distinguish not only between modes vibrating perpendicular from modes vibrating parallel to the surface, but also between parallel modes vibrating parallel (x-modes) or perpendicular (y-modes) to the scattering plane.

To summarize, therefore, two basic experimental methods suffice to establish (a) the applicability of the surface dipole selection rule and (b) the nature of the vibrational modes,

1. the impact energy dependence;
2. the scattering angle dependence; of the vibrational exciatation cross sections;

as portrayed in Figs.6, 15 and 16.

3.6 Hydrogen Modes on W(100)

Experimentally, we expect to see dipole allowed modes close to the specular beam direction and all modes at large scattering angles off specular. Furthermore, the conditions that make the observation of parallel modes most favourable may be summarized as follows:

1. The adsorbate should be located above, rather than within, the metal surface such that its dynamic effective charge is not screened out by local electrodynamic processes. That is, if the adsorbate is located above the image plane, $z \neq 0$, we may represent its position by a point dipole with an effective charge determined by its image dipole below the surface [1].

2. We expect the scattering intensity from each adsorbate atom to depend on the square of the displacement amplitude of that atom for any given normal mode. For a monatomic adsorbate, this displacement is simply given by [8]

$$|u| = (\hbar^2/2M_A\hbar\omega_1)^{\frac{1}{2}} \tag{28}$$

i.e. the adsorbate should possess low mass M_A and the vibrational energy losses $\hbar\omega_1$ should be relatively small. (For a more complicated molecule, this simple relationship between $|u|$ and $\hbar\omega_1$ is no longer true since the displacement of each atom is coupled to its neighbours in a more complicated way).

It is not surprising, therefore, that the first experimental confirmation of these wide angle concepts came from a detailed study of the angle and impact energy dependence of hydrogen adsorbed on a tungsten surface [2].

3.6.1 Angle Dependence

The angular dependence of the loss intensities for a p(1x1)H layer (saturation coverage) on W(100) are shown in Fig.17. Experimentally, we expect to see the dipole allowed symmetric stretch frequency ν_s for H bonded in a C_{2v} point group symmetry bridge site only in the specular direction, $\Delta\theta_s = 0^o$, since this represents vibrational motion normal to the surface [30,31]. Off specular, the vibrational modes parallel to the surface are observed (the asymmetric stretch ν_{as} and the wagging mode ν_w, corresponding to motion along the x and y coordinate directions respectively, Fig.11). Also, in the absence of dipole enhancement effects, the spectral intensities of the parallel modes relate to their vibrational amplitudes, as discussed above. This is seen in Fig.18 which shows the variation in the root-mean-square vibrational amplitudes of the equivalent modes of the triatomic simple harmonic oscillator for which $<u^2> \propto \omega^{-1} \coth (0.72 \omega/T)$, at two temperatures, $T = 125^oK$ and $T = 300^oK$ [32]. The vibrational amplitude decreases with loss frequency indicating the intensity of the low frequency ν_w mode to be higher than that of the higher frequency ν_{as} mode, which is seen to be the case from inspection of the spectral intensities (Fig.17). Attention is drawn to the much weaker overtone and combination bands, the broad widths of which make it difficult to separate out higher harmonic

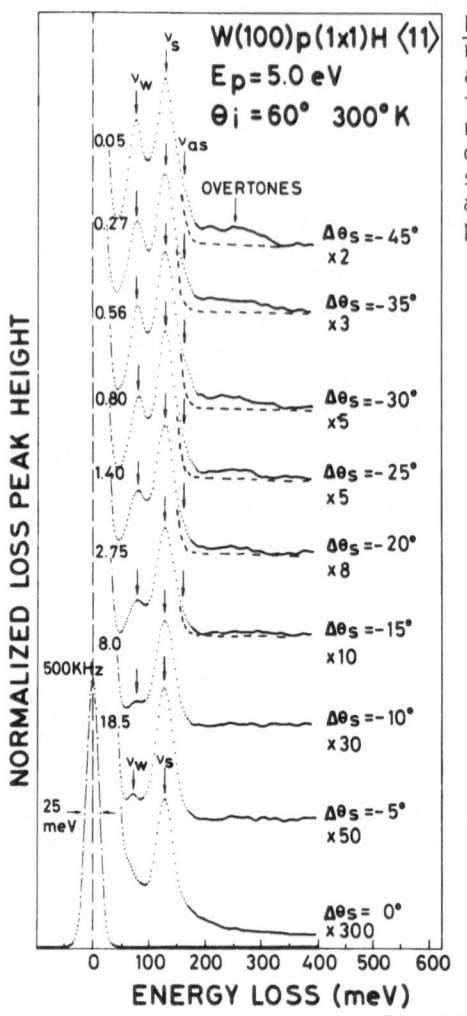

W(100)p(1x1)H ⟨11⟩

$E_p = 5.0$ eV

$\Theta_i = 60°$ 300°K

OVERTONES

ν_w
ν_s
ν_{as}

0.05
0.27 $\Delta\theta_s = -45°$ x2
0.56 $\Delta\theta_s = -35°$ x3
0.80 $\Delta\theta_s = -30°$ x5
1.40 $\Delta\theta_s = -25°$ x5
2.75 $\Delta\theta_s = -20°$ x8
8.0 $\Delta\theta_s = -15°$ x10
500KHz
18.5 $\Delta\theta_s = -10°$ x30
ν_w ν_s
25 meV $\Delta\theta_s = -5°$ x50
$\Delta\theta_s = 0°$ x300

NORMALIZED LOSS PEAK HEIGHT

ENERGY LOSS (meV)

Fig. 17. Angle dependence of the vibrational loss intensities observed in EELS for a p(1x1)H layer on W100) scanning away from the specular beam towards the surface normal ($-\Delta\theta_s^0$) and along the ⟨11⟩ direction of the surface lattice: ν_s, symmetric stretch perpendicular to surface; ν_{as}, asymmetric stretch and ν_w, wagging mode, parallel to the surface.

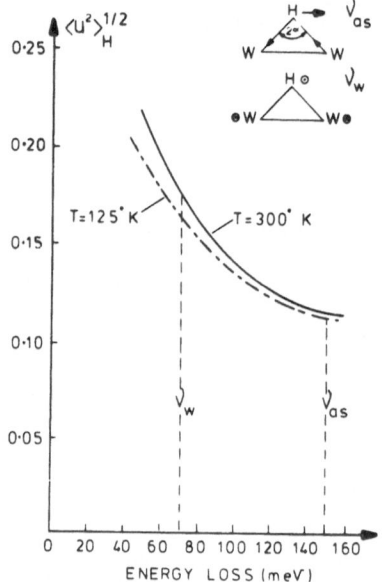

$\langle u^2 \rangle^{1/2}$

0·25
0·20
0·15
0·10
0·05

$T = 125°$ K
$T = 300°$ K

ν_w
ν_{as}

ENERGY LOSS (meV)

Fig. 18. The root-mean-square amplitudes of the vibrational modes of H chemisorbed on W(100) in a C_{2v} bridge site geometry. The modes parallel to the surface, ν_{as} and ν_{wag}, are shown inset.

effects from those due to multiple scattering [33]. The reflection-plane symmetry scattering rule (section 5) would suggest that the observed intensity is mainly a consequence of the combination band ($\nu_{as} + \nu_w$), the scattering due to which will increase off specular in line with the individual fundamental (parallel) vibrational modes [29].

The sensitivity of the scattering cross sections to details of the surface geometry and structure is shown in Fig.19. At saturation coverage ($\beta \sim 1.0$), hydrogen forms a p(1x1) ordered layer in which there are two hydrogen atoms per W(100) unit cell, bridge bonded between two substrate W atoms (re: Fig.11). In the specular direction, the perpendicular mode peaks in intensity due to the long range dipole mechanism, whereas the parallel vibrational modes show a minimum in agreement with the theoretical predictions (re: Figs.10 and 16). Off specular, all modes show a more diffuse

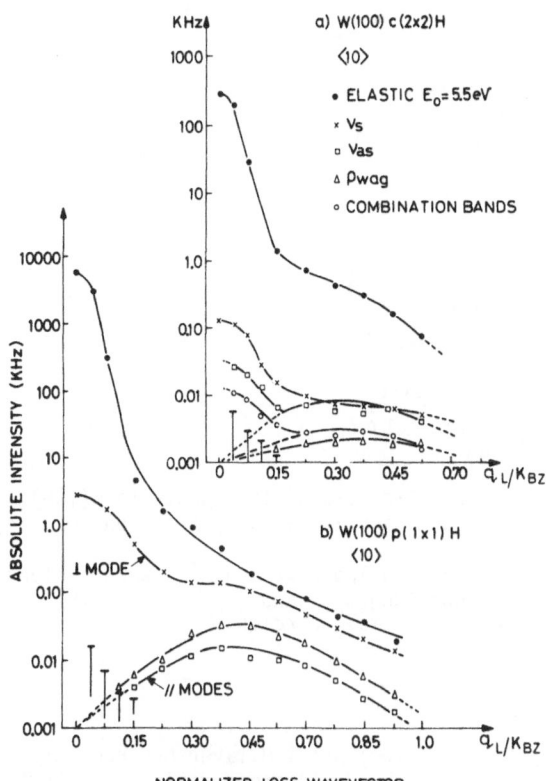

KHz
1000
100
10
1.0
0.10
0.01
0.001

a) W(100) c(2x2)H
⟨10⟩

• ELASTIC E_0=5.5eV
× V_s
□ V_{as}
△ ρwag
○ COMBINATION BANDS

q_L/K_{BZ}

10000
1000
100
10
1.0
0.10
0.01
0.001

ABSOLUTE INTENSITY (KHz)

⊥ MODE

// MODES

b) W(100) p(1x1)H
⟨10⟩

q_L/K_{BZ}

0   0.15   0.30   0.45   0.70   0.85   1.0

NORMALIZED LOSS WAVEVECTOR

Fig.19 Angle dependence of the loss intensities, as a function reduced wavevector (q_L/K_{BZ}), K_{BZ} = 1 referring to the Brillouin zone boundary of
the W(100) square lattice array, for: a) p(1x1)H layer; b) c(2x2)H
reconstructed-surface layer, scanning along the ⟨10⟩ crystallographic
direction. The partial error bars indicate the level of background
noise close to the specular direction (θ_i = 60°) [35].

distribution in intensity, the exact details of which will be dependent on
the operation of the T-matrix in the short range electron impact theory
(section 5). At low coverages ($\beta \sim 0.1$), hydrogen forms a c(2x2) ordered
layer, the structure of which derives from a displacement of pairs of substrate W atoms which come into contact [34]. The hydrogen remains bonded
in the bridge site between two W atoms but now the W-H-W bond angle is
reduced from something of the order of 100° to 80° with a slight elong-
ation of the bond lengths, estimated [35] to be of the order of 5% on the
basis of the structures shown in Fig.20. Also, this shift of the W atoms
involves both a vertical as well as a lateral relative shift such that the
parallel vibrational motions now become tilted up due to a "puckered"
substrate lattice. These modes now possess a vibrational component along
the normal direction which makes them "dipole active" [35]. The effect
of this is seen in the increase in the scattering intensity in the specular
direction for the parallel modes in the c(2x2) structure, absent in the
p(1x1) layer, Fig.19.

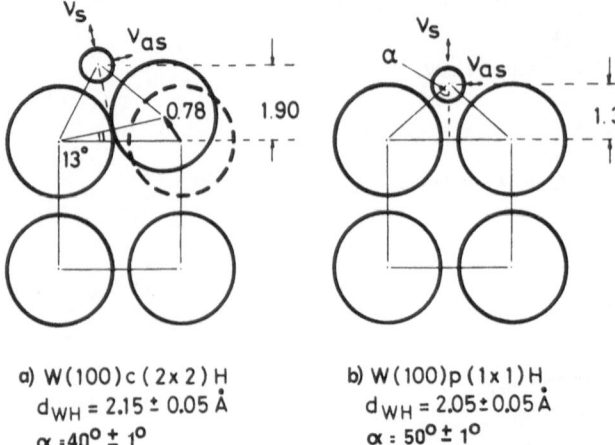

a) W(100)c(2x2)H
$d_{WH} = 2.15 \pm 0.05$ Å
$\alpha = 40° \pm 1°$

b) W(100)p(1x1)H
$d_{WH} = 2.05 \pm 0.05$ Å
$\alpha = 50° \pm 1°$

Fig.20   Showing the magnitude of the displacments of the substrate W(100)
atoms for a) the c(2x2)H reconstructed lattice, in relation to
b) the p(1x1)H surface.
The dimensions refer to Å units assuming W atom displacements along
<11> crystallographic directions [36]. v_a and v_{as} refer to the
vibrational motion of the H atoms (small circles).

3.6.2   Phonon Sidebands Linewidth

The symmetric stretching mode v_s of the W(100) p(1x1)H layer has recently
been observed by an optical technique - surface electromagnetic wave (SEW)
spectroscopy [37]. At room temperature, v_s occurs at 1046 cm^{-1} with a full
width at half maximum of 14 cm^{-1} (8.066 cm^{-1} ≡ 1 meV). This compares with
a width of 118 cm^{-1} (14.7 meV) observed for the equivalent vibrational loss
observed at 1050 cm^{-1} (130 meV) in EELS [30,31]. However, even with the
much poorer resolution of the EELS measurements, this observed width is
more than a factor of two greater than the best instrumental resolution of
6.0 meV (50 cm^{-1}) which has been employed [30,31]. Also, at lower coverages,
the reconstructed c(2x2)H layer produces a shift in the symmetric stretch
frequency to $v_s \sim$ 1250 cm^{-1} (155 meV) [34,35] with a much narrower width,
7.5 meV (60 cm^{-1}). The effective charge of the v_s mode is of the same
order of magnitude in both techniques (SEW, e* ≈ 0.038 e; EELS, e* ≈ 0.04 e).

This marked difference in the v_s spectral line width may be explained by
strong dynamical coupling between the hydrogen vibrational modes and surface
phonons of the substrate lattice. This is illustrated in the recent results
of CHABAL and SIEVERS [36], Fig.21. These authors propose that the EELS
line width is composed of a sharp zero-phonon v_s mode, which is observed
in the optical technique (IR, Fig.21), combined with additional broad
surface-phonon sidebands at higher and lower frequencies corresponding to
one-surface-phonon annihilation and creation processes (D$_+$ and D$_-$ respectively). The absorption coefficient of these side bands is too weak to be
observed by the optical method. However, by combining the two sets of data,
the zero-phonon IR-line may be convolved with the EELS instrumental resolution to produce the dashed curve C (Fig.21) which may then be subtracted
from the EELS spectrum in order to extract the phonon side-band information.

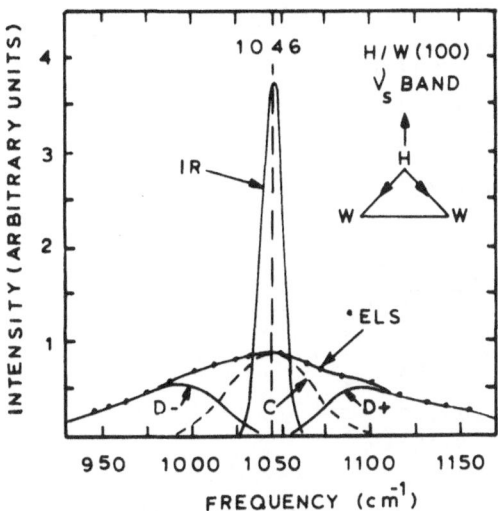

<u>Fig.21</u> The symmetric stretch vibration, ν_S, of p(1x1)H on W(100) as
observed by surface electromagnetic wave (SEW) spectroscopy (curve
IR) [37], and EELS [31]. The phonon sidebands D_+ and D_- are
obtained by subtracting the convoluted IR line (C) from the EELS
data.

This dynamical coupling provides a natural explanation of the H/W(100)
EELS line widths. Since a saturation coverage of hydrogen causes a static
rearrangement of the W surface atoms back to a cubic W(100) lattice, strong
coupling between the hydrogen vibrations and the surface phonons is implied.
In the case of the reconstructed c(2x2)H layer, a simple geometric argument
[34,35] shows that the ν_S frequency will decrease as the W-H-W bond angle
decreases from $\sim 100^\circ$ to $\sim 80^\circ$. This implies weaker static coupling which,
in turn, implies weaker dynamic coupling so that most of the intensity in
the low coverage ν_S mode will remain in the zero phonon transition, producing a much narrower line width, as is observed experimentally [30,31].

This example emphasizes the importance of the degree to which the
adsorbate A and substrate S phonon modes couple together and determine the
overall line width of the vibrational excitation.

3.6.3 Impact Energy Dependence

The small-angle dipole scattering theory [1] suggests that it is useful to
plot the ratio $I_1(E)/I_0(E)$, where $I_1(E)$ and $I_0(E)$ are the one-phonon and
(elastic) specular beam intensities respectively. Any deviation from a
smoothly varying function of impact energy can be attributed to rapid
variations in the reflectivity and phase of the incident electrons, as
discussed in section 3.2. However, off specular in the large scattering
angle regime, rapid fluctuations in both the elastic and inelastic signals
can occur due to the combined effect of multiple elastic scattering and
summing over the inelastic scattering amplitude derivatives in the vibrational matrix elements (section 5). Unfortunately few experimental measurements have been made to date due to the instrumental difficulty of maintain-
ing constant transmission through the spectrometer as the monochrometer and
analyzer pass energies are simultaneously swept.

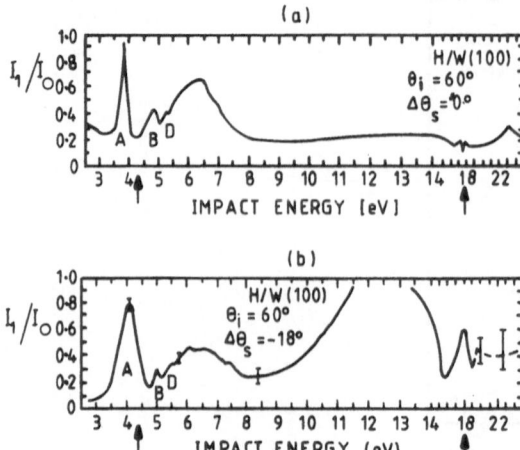

<u>Fig.22</u> Energy dependence of the normalized loss intensity, I_1/I_0, for the perpendicular mode, ν_s, of a p(1x1)H layer on W(100) [5,22]
a) specular beam direction; b) $\Delta\theta_s = -18^0$ away from specular, towards the surface normal. The arrows indicate the emergence energies of the $0\bar{1}$ and $0\bar{2}$ diffraction beam; the associated structure A, B, D, etc. is shown in more detail (Fig.7 and text).

A first attempt at such measurements over a limited range of impact energies, 2 eV $\simeq E_0 \simeq$ 25 eV, is shown in Fig.22 for scattering from the ν_s mode of the p(1x1) layer of H on W(100) [5,22]. Here we compare the impact energy dependence of the ratio (I_1/I_0) for the two extremes of scattering, in the specular direction (Fig.22a) with scattering $\Delta\theta_s \simeq -18^0$ off specular (Fig.22b), as defined by the scattering geometry, Fig.9. The normalized loss intensity shows considerable structure. Resonance fluctuations A, B, D, etc. occur at the thresholds associated with the grazing emergence of diffraction beams (arrowed) due to short range scattering associated with the bound state surface resonances described earlier (section 3.2), which persist in both the specular and off specular curves. In addition, much broader features occur, the origin of which is presumably due to multiple scattering effect (section 5) since these features are more pronounced in the off specular data (Fig.22b).

As outlined in section 5, the impact energy dependence of the EELS vibrational cross sections is sensitive to the detailed structure of the adsorbate bonding sites [10]. However, in the large-angle regime it is not appropriate to plot the normalized loss intensities (I_1/I_0) since the variation in the elastic $I_0(E)$ and inelastic $I_1(E)$ cross sections are not necessarily related. This is illustrated with results taken from TONG, LI and MILLS [27], Fig.23, in which we see structure in the phonon loss cross section to be distinctly different from that in $I_0(E)$, the latter being the specular beam I-V curve (dashed curve) familiar in conventional LEED measurements [26]. It is therefore more useful to record $I_1(E)$ directly in the case of off specular impact scattering for comparison with the theoretical predictions [27].

50

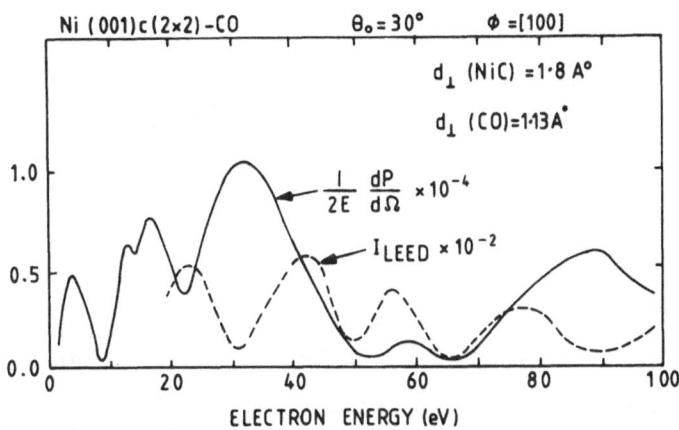

Fig.23 Energy dependence of elastic specular intensity I_0 (broken line)
        and inelastic loss intensity I_1 (full line) for the C-O stretch
        vibration (ω_+), Fig.15 [27].

3.7 Concluding Remarks

The above experimental results would suggest that our present picture of
the mechanism of inelastic scattering of electrons from adsorbates on metal
surfaces is reasonably complete. The results for CO on nickel and copper
surfaces (section 3) serve to verify the long range dipole interaction
theories. The H/W(100) EELS spectra (section 6) endorse the view that the
angular dependence of the inelastic scattering cross sections easily distinguishes between those modes vibrating parallel from those modes vibrating
normal to the surface. However, it is obviously desirable to formulate a
single complete theory of EELS which would include both the small-angle
dipole-excitation and impact-scattering mechanisms. Also, several problems
remain.

    The distinctive angular dependence of the hydrogen modes in W(100) is
not observed on other metal surfaces. For example, H/Ni(111) shows two
loss peaks, the intensities of which remain more or less constant with
scattering angle away from the specular directions [5]. One suggestion in
this case [17] is that the hydrogen lies much deeper in the surface in a
3-fold coordination site such that it is more effectively screened by the
surrounding metal [1]. On the other hand, similar behaviour is observed
for C$_2$H$_2$ adsorbed on Ni(100), which is not so easily "buried" in the
surface. It may well be that these anomalies reflect differences in the
adsorbate-surface electrodynamics due to differences between the different
chemisorptive bonds and the electronic polarizabilities, yet to be understood
[1]. Also, an inherent assumption in most theories is that of one-phonon
excitation processes; it would be useful to understand any effects due to
disorder on the angle and energy dependent behaviour of the scattering cross
sections. A related topic is that of phonon dispersion effects due to coupling between the adsorbate atoms (treated in the following chapter by
DOBRZYNSKI). The problem of the collective vibrational modes of adsorbates
and surfaces is one for the future, in which electron scattering spectroscopy
is sure to play a significant role.

Acknowledgements

In writing this article, I have benefited greatly from access to the unpublished work of many people, especially that of S. Andersson, C. Backx, W. Ho, H. Ibach, B.N.J. Persson, E.W. Plummer. J.H. Rose, S.Y. Tong, N.V. Richardson, K.L. Sebastian and J.W. Wilkins. I would also like to thank the Royal Society (London) for support for this work.

References

1. D.M. Newns, Chap.2, this volume.

2. W. Ho, R.F. Willis, E.W. Plummer; Phys. Rev. Lett., $\underline{40}$, 1463 (1978).

3. For a recent extensive review see, W.H. Weinberg in "Experimental Methods of Surface Physics", ed. R.L. Park (Academic Press, N. York) 1980.

4. J.W. Davenport, W. Ho, J.R. Schrieffer; Phys. Rev., $\underline{B17}$, 3115 (1978).

5. W. Ho, Ph.D. Dissertation, University of Pennsylvania, Philadelphia (1979).

6. See review article by G.J. Schulz, Rev. Mod. Phys., $\underline{45}$, 423 (1973).

7. Z. Lenac, M. Sunjic, D. Sokcevik, B. Brako; Surf. Sci., $\underline{80}$, 602 (1979); Z. Physik, $\underline{B28}$, 273 (1977).

8. J.E. Rose, J.W. Wilkins; unpublished.

9. G.E. Thomas, W.H. Weinberg; J. Chem. Phys., $\underline{70}$, 1000 (1979).

10. C.H. Li, S.Y. Tong, D.L. Mills; Phys. Rev., $\underline{B21}$, 3057 (1980).

11. K.H. Johnson, Adv. Quantum Chem., $\underline{7}$, 143 (1973).

12. D. Dill, J.L. Dehmer; J. Chem. Phys., $\underline{61}$, 692 (1974).

13. E. Evans, D.L. Mills; Phys. Rev., B5, 4126 (1972); D.L. Mills in Progress in Surf. Sci., $\underline{8}$, 143 (1977).

14. D.M. Newns, Phys. Lett., $\underline{60A}$, 461 (1977).

15. B.N.J. Persson, Solid State Commun., $\underline{24}$ 573 (1977).

16. F. Delanaye, A.A. Lucas, G.D. Mahan; Surf. Sci., $\underline{70}$, 629 (1978).

17. E.W. Plummer, W. Ho, S. Andersson; to be published.

18. S. Andersson, B.N.J. Persson, T. Gustafsson, E.W. Plummer; Sol. State Commun., $\underline{34}$, 473 (1980).

19. B.N.J. Persson, Surf. Sci., $\underline{92}$, 265 (1980).

20. G. Blyholder, J. Phys. Chem., $\underline{68}$, 2772 (1964); ibid, $\underline{19}$, 756 (1974).

21. S. Andersson, J.W. Davenport; Sol. State Commun., $\underline{28}$, 677 (1978).

22. R.F. Willis, W. Ho, E.W. Plummer; Surf. Sci., $\underline{80}$, 593 (1979); Phys. Rev. B, in press (1980).

23. A. Adnot, J.D. Carette; Phys. Rev. Lett., $\underline{39}$, 209 (1977).

24. R.F. Willis; Phys. Rev., $\underline{B17}$, 909 (1977).

25. H.A. Pearce, N. Sheppard; Surf. Sci., $\underline{59}$, 205 (1976).

26. S.Y. Tong; Progress in Surf. Sci., $\underline{7}$, 1 (1975).

27. S.Y. Tong, C.H. Li, D.L. Mills; Phys. Rev. Lett., $\underline{44}$, 407 (1980).

28. J.W. Davenport; Phys. Rev. Lett., $\underline{36}$, 945 (1976).

29. K.L. Sebastian; Phys. Rev. Lett., $\underline{44}$, 1414 (1980); J. Phys. C: Solid State Phys., $\underline{13}$, L115 (1980).

30. H. Froitzheim, H. Ibach, S. Lehwald; Phys. Rev. Lett., $\underline{36}$, 1549 (1976).

31. A. Adnot, J.D. Carette; Phys. Rev. Lett., $\underline{39}$, 209 (1977).

32. S.J. Cyvin, "Molecular Vibrations and Mean Square Vibrational Amplitudes", Elsevier, Amsterdam (1968); N.V. Richardson, Private communication.

33. Compare the results, Ref. 21.

34. M.R. Barnes, R.F. Willis; Phys. Rev. Lett., $\underline{41}$, 1729 (1978).

35. R.F. Willis, Surf. Sci., $\underline{89}$, 457 (1979).

36. M.K. Debe and D.A. King; Phys. Rev. Lett., $\underline{39}$, 708 (1977).

37. Y.J. Chabal, A.J. Sievers; Phys. Rev. Lett., $\underline{44}$, 944 (1980).

4. Adsorbate Induced Optical Phonons

L. Dobrzynski, G. Allan, B. Djafari-Rouhani, and B. K. Agrawal[1]

With 4 Figures

Abstract

An adsorbed monolayer modifies the surface phonons and may create new localized or resonant modes. These main physical effects are discussed, mainly for
optical phonons.

Then this lecture is illustrated by a review of the theoretical results
obtained for the optical phonons due to hydrogen on the (001) surface of
tungsten and to oxygen on the (111) surface of nickel. These results are compared to the available experimental data.

4.1. Background

Adsorbed atoms modify many vibrational properties of surfaces and create new
ones [1]. The adsorption entropies can be obtained by measuring the desorption rate of the adsorbed particles as a function of temperature or by anal-
yzing the adsorption isotherms determined by Auger spectroscopy. The mean
square displacements of adsorbed atoms can be measured in L.E.E.D. and Möss-
bauer experiments. The low temperature specific heat, the phonon free energies, the local phonon density of states are also modified by adsorption. The
elastic and phonon contributions to the free energy of interacting adatoms
pairs are also essential. The localized and resonant modes of vibration due
to adsorption are studied as well in the acoustical region [1,2] as in the
optical one.

Our aim here is not to review and discuss all these properties [1], we
will stick to the main interest of this Symposium and focus on adsorbate induced optical phonons.

We first review the theoretical works done on this subject and discuss the
main physical effects. Then we illustrate this lecture by going in more details through the theoretical studies of the optical phonons due to hydrogen
on tungsten (001) and then to oxygen on nickel (111), comparing them with
the experimental results. The choice of these two examples is somewhat arbitrary and due only to the fact we know them better than the others.

4.2 A Short Review

The existence of absorbate induced optical phonons was first shown on two-
dimensional lattice models. Kaplan [3] and Hori et al [4] put a row of isotopic impurities along one of the free edges of a two-dimensional lattice.

---

1 Permanent address: Department of Physics, Allahabad University, Allahabad
  India.

They found that new surface modes can appear whose frequencies are higher than the frequencies of the bulk crystal phonons for the same wave vector.

Adsorbate induced optical phonons due to an absorbed layer of impurity atoms on the surface of a three-dimensional crystal were then obtained by DOBRZYNSKI et al [5]. They used a simple cubic crystal with nearest neighbour, central and noncentral forces, and assured that the atoms in the absorbed layer, deposited on a (001) free surface, differ from those of the substrate only in their masses.

The localized modes of vibration due to the adsorption of a monolayer were then calculated within more and more realistic models. DOBRZYNSKI [6] and ARMAND et al [7] studied (001) surfaces of b.c.c crystals with application to H on W in particular. MASRI et al [8] used for a qualitative discussion of the different types of modes due to adsorption a still analytical model of a simple cubic crystal with central forces between pairs of nearest and second nearest neighbour atoms. Rare gas monolayers on another rare gas b.c.c solid bounded by (111) and (100) surfaces were investigated by ALLDREDGE et al [9], LAWRENCE et al [10] and CASTIEL et al [11]. LAWRENCE et al [10] took into account the effect of static relaxation of the distance between substrate and adsorbate.

The effects of a superstructure on the optical surface modes was first described by DOBRZYNSKI et al [12] for a (2x1) monolayer superstructure on a (001) surface of a simple cubic crystal. Realistic calculations including these effects appeared for monolayers of: S and O on (001) and (111) surfaces of Ni, by ARMAND et al [13] and ALLAN et al [14]; H on (001) surfaces of W, by BULLETT et al [15], AGRAWAL et al [16] and RICHARDSON et al [17]; Xe on (001) graphite by COULOMB et al [18] and for Ar and N monolayers on graphite by GILLIS et al [19].

Models deduced from the surface electronic band structure were studied by ALLAN et al [14] for O and S on Ni surfaces and BULLETT et al [15] for H on W. ALLAN et al [14] are able to fit with the same set of parameters the experimental results for the distances between adsorbate and substrate as well as the phonon frequencies.

Let us mention finally a simple theory of the variation with temperature of the vibrational modes of an adsorbed monolayer. The temperature-dependent relaxations of the interplanar spacings were assumed by DJAFARI-ROUHANI et al [20] to arise only from cubic anharmonic terms in the crystal potential energy. The change in force constants is then obtained and finally the variation with temperature of the surface phonons. However quartic anharmonic terms are expected to reduce the magnitude of the variations obtained in this work with the cubic anharmonic terms only.

4.3 Some Physical Effects

Let us now discuss the main qualitative effects displayed in all these studies.

Adsorbed atoms lighter than the substrate atoms increase the frequencies of the surface modes and new modes may also appear above the bulk phonon bands for a given value of the propagation vector $\vec{k}$ parallel to the surface.

An increase in the surface force constants has a similar effect to a light monolayer. Such an increase in the force constants can in particular exist when the distances between the surface and the adsorbed atoms are contracted.

There can be "window gaps" within the sub-bands corresponding to acoustic bulk modes (see Fig.1). High-frequency localized modes can be, and usually are, present within these gaps (part of curve (2) on Fig.1).

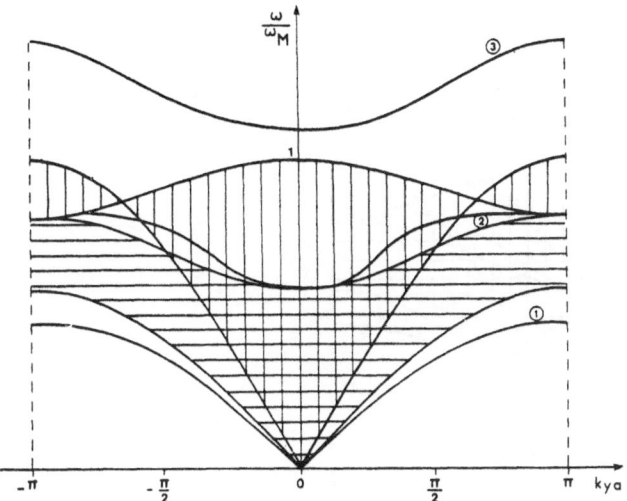

Fig.1 Different types of surfaces waves due to adsorption of a monolayer in registry with the substrate. The shaded areas give the bulk bands. Curve (1) represents the Rayleigh waves modified by adsorption. Curve (2) corresponds to localized waves falling partly into a window gap and partly into a bulk band where the atom displacements are orthogonal to those of this mode. Curve (3) shows an optical phonon branch situated above the bulk bands.

It is possible for surface modes to exist even within some bulk sub-bands. For example, the curve (2) is due to a localized mode polarized along the axis orthogonal to the propagation vector k_y and the normal $\hat{z}$ to the surface. This curve falls partly in a window gap and partly for smaller value of k_y within the bulk sub-band corresponding to phonons polarized within the (y0z) plane. Therefore these surface modes can remain decoupled from the bulk phonons having the same frequencies.

Let us mention also that one can have localized surface phonons in one part of the two-dimensional Brillouin zone. Usually such a branch of localized modes can be prolongated by a branch of resonances, which may correspond to a sharp feature in the local phonon density of states. These effects were studied in detail by MASRI et al [21]. In particular adsorption may cause that one part of a surface phonon branch falls in the bulk continuum or inversely that localized modes peel off the bulk bands only in a part of the two-dimensional surface Brillouin zone.

In many instances, the atoms within an adsorbed monolayer assume a configuration with symmetry lower than that associated with the surface of the substrate. One finds [12] that a superstructure has two main qualitative effects on the properties of the losalized phonons. These effects are illustrated in a schematic fashion on Figs.1 and 2.

Consider first the case where the surface has no superstructure and examine the vibrational spectrum of the semi-infinite crystal with an adsorbed layer having the same symmetry as the (001) surface. This vibrational spectrum is sketched on Fig.1 for modes with wave vector $\vec{k}$ directed along the y direction, $k = 0$. For each value of k_y one has a range of frequencies asso-

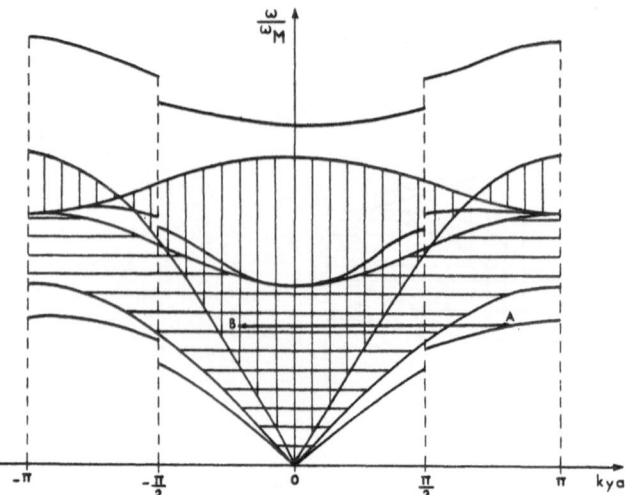

Fig.2 Effect of a (2x1) superstructure in the adsorbed monolayer on the surface phonons given on Fig.1.

ciated with bulk vibrations (shaded areas). The width of these bulk bands for
each value of k_y is due to the fact that for each value of $\vec{k}$ one has several
values of $\omega(\vec{k},k_z)$ as a function of k_z which is still a good quantum number
for bulk bands.

On Fig.1, we assumed the existence of three branches of surface modes, related to the adsorption of a monolayer having the same symmetry as the substrate. Curve (1) represents the Rayleigh waves modified by adsorption. Curve
(2) shows a branch of modes peeled off the bulk phonon band (horizontally
shaded area) situated just below and corresponding to atom displacements along
the $\hat{x}$ axis. This localized mode branch (2) falls partly into a "window gap"
and partly into a bulk phonon band (vertically shaded area) where the atom
displacements lie in the (yOz) plane. Note that here the third bulk phonon
band, corresponding also to atom displacements in the (yOz) plane corresponds
to the same frequencies as the bulk band with polarisation x (horizontally
shaded area). Curve (3) shows another adsorbate induced optical phonon branch
corresponding to atom displacements in the sagittal (yOz) plane.

Now suppose the adsorbed monolayer has a superstructure, in such a manner
that the new repeat distance parallel to the $\hat{y}$ axis assumes a value twice
that appropriate to the (001) surface of the substrate. We can plot the normal mode spectrum in this case as indicated in Fig.2.

The effect of the superstructure is to introduce new zone boundaries at
$k_y = \pm\pi/2a$, respectively where a is the lattice parameter. The region
$-\pi/2a < k_y < \pi/2a$ is now the first Brillouin zone and the two regions
$\pi/2a < k_y < \pi/a$, and $-\pi/a < k_y < -\pi/2a$ correspond to the second Brillouin zone.
Gaps in the surface mode dispersion relations open up at the zone boundaries,
as indicated on Fig.2. If the surface phonon dispersion relations are plotted
in the reduced zone ($-\pi/2a < k_y < \pi/2a$), there are now twice as many distinct
surfaces branches (six in this example).

The surface modes which were at $k_y = \pi/a$ for the unreconstructed layer, are now at $k = 0$ and can be detected by the usual spectroscopic experiments. In particular the acoustic surface mode at $k_y = \pi/a$ is now detectable by the high frequency spectroscopic methods [22].

There is one other effect that is quite striking. Consider the surface mode at point A in Fig.2. This mode is well localized for the monolayer having the same symmetry as the surface. When the monolayer has its (2x1) superstructure, the resulting perturbation term in the dynamical matrix mixes this surface phonon A with the bulk phonon B displaced in $\vec{k}$ from A by the reciprocal lattice vector $\vec{G} = -\hat{y}\pi/a$. As a result of this admixture the mode A is no longer a true localized mode, but becomes a virtual state (called also resonance), since the displacement field is no longer localized at the surface. In general, for a localized surface phonon to be converted to a virtual state by this process, its frequency must fall into a bulk band when one folds back all the bands into the reduced Brilluoin zone. We will now illustrate this general discussion on two specific examples for which experimental results are available.

4.4 Hydrogen on (001) Surface of Tungsten

Hydrogen being the lightest atom that could adsorb on a surface is of particular interest. Vibrational frequency data have been obtained from infrared transmittance experiments, reflection-absorption infrared, electron energy loss and inelastic neutron scattering techniques. We do not review here all these results obtained for different crystal surfaces, as this was done by JAYASOORIYA et al [23] but focus on those [24-28] obtained for the (001) surface of tungsten, by electron energy loss spectroscopy. We will namely try to interpret the recent results obtained by WILLIS et al [26-28].

The clean W(001) surface is unstable below 300 K and it reconstructs into a C(2x2) superstructure. However, the C(2x2) superstructure can be stabilised above 300 K by the chemisorption of hydrogen. One finds a distinct LEED pattern corresponding to a C(2x2) superstructure namely β_2-phase. Further increased coverage of H drives the C(2x2) superstructure back to the normal p(1x1) surface lattice similar to a (001) plane in the bulk W. A distinct LEED pattern corresponding to p(1x1) lattice develops fully for a saturation coverage of H corresponding to a monolayer coverage of β_1 phase.

A model for the atomic positions for the C(2x2) H structure for low coverage of H seems to be still open [16,23,28]. We will not discuss it here. We will just stick to the simple β_1 phase and show that even in this simpler case, we were not able to decide about a definitive dynamical model for the vibrations of H on W(001).

The first difficulty in all theoretical studies of adsorbate induced optical phonons is to know the atomic position of the atoms. If these positions are not known, it may be possible to fit the experimental vibrational frequencies for different possible atomic positions. This was demonstrated in particular for the reconstructed β_2 phase of H on W(001) [16]. Even when there is no reconstruction like in the β_1 phase, the distance between the H monolayer and the surface of the W crystal remains unknown. In what follows, we will present a lattice dynamical model for this system and will show that this model is not the unique one which may fit the three measured vibrational frequencies.

We refer to Fig.3 for the geometry of the unreconstructed W(001) surface with H-atoms. We have two H-atoms per square planar unit cell of lattice parameter a, as an H atom is bound in a bridge position to each pair of W atoms situated on the (001) surface.

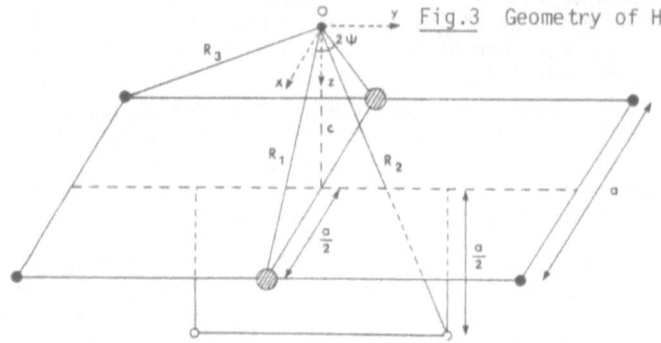

Fig.3 Geometry of H on W(001) in the β_1 phase

•   H

◉   First neighbour W-atoms of H

○   Second neighbour W-atoms of H

●   Third neighbour W-atoms of H

We assume three central pair potentials (ϕ_1, ϕ_2, ϕ_3) between a H-atom and its first, second and third nearest neighbours. For the adatom-adatom inter-actions, i.e., H-H interactions, we assume a single central force constant β. We neglect the angular interactions for the W-H-W and H-W-W angles.

As the H-atom is much lighter (1/180) compared to the W-atom, its frequencies lie much above the W-bulk phonon bands and one may work in the frozen substrate approximation, i.e., all the W-atoms are frozen.

In a set of axes where the $\hat{x}$-axis lies along the W-W line, the $\hat{y}$-axis normal to it, and the $\hat{z}$-axis points downwards normal to the surface (see Fig.2), the three Fourier-transformed equations of motion for a H-atom of mass m can be written as

$$(-m\omega^2+4\beta+K_x)\vec{e}_x(1) = 4\beta \cos(k_x a/2)\cos(k_y a/2)\vec{e}_x(2)$$
$$- 4\beta \sin(k_x a/2)\sin(k_y a/2)\vec{e}_y(2) \tag{1a}$$

$$(-m\omega^2+4\beta+K_y)e_y(1) = -4\beta \sin(k_x a/2)\sin(k_y a/2)\vec{e}_x(2)$$
$$+ 4\beta \cos(k_x a/2)\cos(k_y a/2)\vec{e}_y(2) \tag{1b}$$

$$(-m\omega^2+K_z)\vec{e}_z(1) = 0 \tag{1c}$$

where $\vec{e}(1)$ and $\vec{e}(2)$ denote the eigenvectors of the two H-atoms of a planar unit cell; ω is the phonon frequency corresponding to the wave vector $\vec{k}$ and the three diagonal force constants are given by

$$K_x = 4\left(\frac{a}{R_3}\right)^2\phi''(R_3) + \frac{1}{2}\left(\frac{a}{R_2}\right)^2\phi''(R_2) \quad , \tag{2a}$$

$$K_y = \frac{1}{2}\left(\frac{a}{R_1}\right)^2\phi''(R_1) + \left(\frac{a}{2R_3}\right)^2\phi''(R_3) \quad , \tag{2b}$$

$$K_z = 2\left(\frac{c}{R_1}\right)^2\phi''(R_1) + 2\left(\frac{c+a/2}{R_2}\right)^2\phi''(R_2) + 4\left(\frac{c}{R_3}\right)\phi''(R_3) \quad , \tag{2c}$$

where $\phi''(R)$'s are the second derivatives of the three pair potentials for first, second and third W-atom neighbours of a H-atom. R_1, R_2 and R_3 denote the distances of the neighbours explicitly given by

$$R_1 = (c^2+a^2/4)^{\frac{1}{2}} \quad ; \quad R_2 = [(c+a/2)^2+a^2/4]^{\frac{1}{2}} \quad ; \quad R_3 = (c^2+5a^2/4)^{\frac{1}{2}} \tag{3}$$

with c as the monolayer-substrate surface separation.

Equation (1c) gives the frequency of the non-dispersive mode having displacements perpendicular to the surface as

$$m\omega_z^2 = K_z \quad . \tag{4}$$

Equations similar to (1a) and (1b) can easily be written down for the second H-atom of the unit cell by interchanging the value of $\kappa(=1,2)$ and also of $\alpha(=x,y)$ in the eigenvectors $e_\alpha(\kappa)$ and further, the α indices in the k_α and K_α.

The frequencies of the modes polarized parallel to the surface are then obtained from Eqs.(1a) and (1b) and their transposed ones. The four eigenfrequencies of the (4x4) dynamical matrix are thus given by

$$m\omega_{1,2}^2 = \frac{1}{2}(K_x+K_y) + 4\beta[1-\sin(k_xa/2)\sin(k_ya/2)]$$
$$\pm \frac{1}{2}\left[(K_x-K_y)^2+64\beta^2\cos(k_xa/2)\cos(k_ya/2)\right]^{\frac{1}{2}} \quad , \tag{5a}$$

$$m\omega_{3,4}^2 = \frac{1}{2}(K_x+K_y) + 4\beta[1+\sin(k_xa/2)\sin(k_ya/2)]$$
$$\pm \frac{1}{2}\left[(K_x-K_y)^2+64\beta^2\cos(k_xa/2)\cos(k_ya/2)\right]^{\frac{1}{2}} \quad . \tag{5b}$$

The corresponding two-dimensional Brillouin zone is a square in which

$$-\pi/a < k_x < \pi/a \quad \text{and} \quad -\pi/a < k_y < \pi/a \quad .$$

It may be noted that the dispersion in all the H-modes is only due to the H-H interactions β. However, this dispersion has not been detected in experiments [26-28] because of the small observed widths ($\sim$5 meV) of the H-atom bands. One may infer that the adatom-adatom interaction is quite weak and we take $\beta = 0$ in further calculations.

Then for $k_x = k_y = 0$, we have the following two doubly-degenerate different frequencies

$$m\omega_1^2 = m\omega_3^2 = K_x \quad , \tag{6a}$$
$$m\omega_2^2 = m\omega_4^2 = K_y \quad . \tag{6b}$$

As the distance c between the H monolayer and the (001) W surface is not yet precisely known, we need some way for estimating it. For this purpose we work in a nearest–neighbour approximation where the two frequencies ω_x and ω_z which correspond to the parallel and perpendicular polarizations of the modes lying in the plane of the molecule W-H-W are related as

$$\frac{\omega_z^2}{\omega_x^2} = \frac{K_z}{K_x} = \frac{4c^2}{a^2} = \cot^2\psi \quad . \tag{7}$$

By employing the two high experimental frequencies 160 meV and 130 meV for

ω_z and ω_x [25-27] respectively, one estimates $\psi \simeq 51°$ and with $a = 3.16$ Å, one obtains $c = 1.28$ Å and $R_1 = 2.03$ Å.

Let us note that this simple relation (7) was found [23] to hold for a range of cluster transition metal complexes with adsorbed H in bridged positions and was used to predict [28] the type of reconstruction which should occur in the β_2 phase.

We see here that first nearest neighbour approximation, is not able to explain the third observed frequency at 80 meV, because $K_y = 0$ within this approximation. When keeping the interactions up to the third nearest neighbour, it is possible [16] to fit the three experimental frequencies, with the following values of the second derivatives of the potentials:

$$\frac{\hbar}{m} \phi''(R_1) = 2.02 \times 10^4 \text{ meV}^2 \quad , \quad R_1 = 2.03 \text{ Å}$$

$$\frac{\hbar}{m} \phi''(R_2) = -0.0075 \times 10^4 \text{ meV}^2 \quad , \quad R_2 = 3.27 \text{ Å}$$

$$\frac{\hbar}{m} \phi''(R_3) = 0.24 \times 10^4 \text{ meV}^2 \quad , \quad R_3 = 3.76 \text{ Å} \quad . \tag{8}$$

However as mentioned before, the above model is not unique. When one takes into account angular interactions, one gets another parameter. When neglecting one of these four parameters, in order to fit the three experimental frequencies. AGRAWAL et al [16] obtained three different models, which were found all plausible.

4.5 Oxygen on the (111) Surface of Nickel

The preceding example shows that the surface-adatom distance is needed, as well as the adsorption geometry, before any quantitative fit of the adsorbate induced optical phonons can be made. This distance is for some systems known from LEED measurements, otherwise one may approximate it from molecular interatomic distances or by some simpler procedure as done above. Therefore a model which could give the vibration frequencies as well as the interatomic distances is of great interest. ALLAN et al [14] proposed recently such a model starting from surface electronic structure calculations.

They describe the transition metal substrate d states and the oxygen p states within the tight-binding approximation. They take with a good approximation each atom neutral. Only two first moments of the local density of states are used to calculate the electronic energies. The local density of states on a given atom are fitted to a Gaussian.

The hooping integrals β_{ij} between atoms i and j depend on the distance R_{ij}

$$\beta_{ij} = \beta°_{ij} \exp(-q_{ij} R_{ij}) \quad . \tag{9}$$

The stability of the crystal is ensured by a Born-Mayer repulsive potential

$$C_{ij} = C°_{ij} \exp(-p_{ij} R_{ij}) \quad . \tag{10}$$

The $\beta°_{ij}$, q_{ij}, $C°_{ij}$ and p_{ij} are parameters of this model.

A first and second order development of the energy as a function of atomic positions is used to minimize the energy, find the interactomic distances and the dynamical matrix. The eight parameters of this model are fitted to bulk transition metal properties, to molecular vibration frequencies or to the observed electron loss spectra.

Recent electron loss spectroscopy measurements of O chemisorbed on a Ni (111) surface show peaks which were assigned by IBACH et al [22] to Ni surface phonons. Electron losses are only observed at the center of the surface Brillouin zone. However, two oxygen structures ($\sqrt{3} \times \sqrt{3}$ R 30°) and C(2x2) are known. As explained in Section 3, the new unit cell in real space is bigger after adsorption and the new two-dimensional Brillouin zone smaller. Thanks to the so produced folding of the phonon bands, localized clear surface phonons recently calculated by VELASCO et al [29] and situated at the edge of the Brillouin zone can now be detected for k = 0. If the oxygen adatoms do not too much perturb the free surface vibrations, these localized vibrations may induce energy losses via the dipole due to charge transfer between the oxygen and its nickel neighbours.

O on Ni is one of the best known systems. Its crystallography seems to be well determined by LEED, for (001), (111) and (110) surfaces. The vibration frequencies for displacements perpendicular to the surfaces were also measured. ALLAN et al [14] fitted with their model all the experimental data with good accuracy, apart for the distance O-Ni at the (001) surface and the corresponding frequency for the C(2x2) structure. Note that this system is worth complementary studies from the experimental point of view as well as from the determination of the atomic positions. We will show here only the results for the p(2x2) structure on the (111) surface.

The phonon spectra were calculated at the center of the structure Brillouin zone using a slab of 6 to 12 Ni planes deposited on a frozen Ni substrate. The local phonon frequencies have been Gaussian broadened to get a continuous spectrum. Figure 4 gives the spectra calculated for the (111)p(2x2) structure. The vibrating dipole $D(\omega)$ is represented on curve (A)

$$D(\omega) = \sum_n \sum_i Q_i |u_i(\omega_n)| \delta(\omega - \omega_n) \tag{11}$$

where $Q_i = 1$ for the oxygen atom and $Q_i = -1/N$ for one of its N nickel neighbours and $u_i(\omega_n)$ is the amplitude of vibration on atom i for the frequency ω_n. It is slightly different from the local density of phonons which is

$$n_i(\omega) = \sum_n |u_i(\omega_n)|^2 \delta(\omega - \omega_n) \quad . \tag{12}$$

Curves (B), (C) and (D) give, respectively, the local phonon spectra on the oxygen adatom, on one of its nickel neighbours and on a nickel free surface atom.

The peaks of $D(\omega)$ agree quite well with the experimental ones. Note that ALLAN et al [14] did not compare the scattering amplitudes but only the peak positions as they have only calculated the vibrating dipole and not scattering probability.

One can see on the curves (C) and (D) that the adsorption of oxygen modifies the local nickel surface atom phonon spectra. The localized surface vibration frequencies are slightly shifted. New force constants appear between the surface atoms and the oxygen atoms. Moreover, the oxygen adsorption destroys the contraction arising near a free nickel surface or even may change the relaxation sign. This variation of the relaxation also modifies the force constants between atoms close to the surface and then the surface vibration frequencies.

Note that the dotted lines on Fig.4 are for vibrations polarized in the surface plane. These theoretically predicted peaks should be observed in forthcoming experiments, similar to the one done for H on W(001) [26-28].

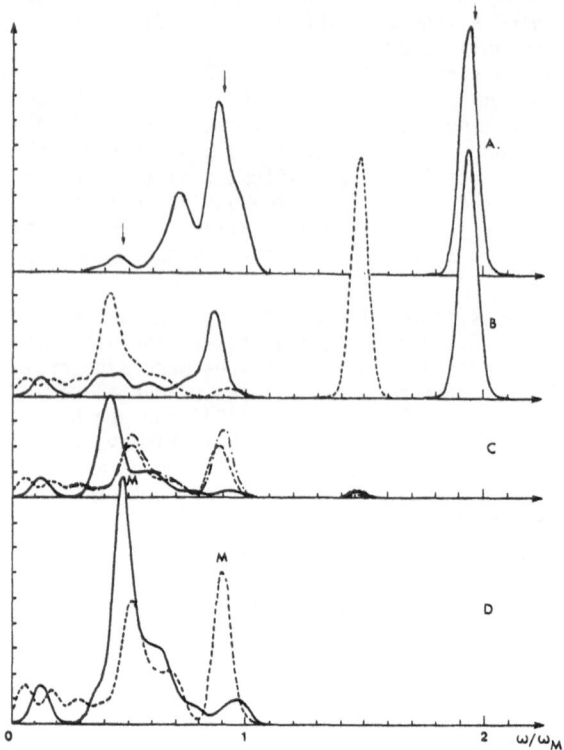

Fig.4 Oxygen p(2x2) structure local phonon spectra on Ni(111) surface. The curve (A) shows the vibrating dipole. The curves (B), (C) and (D) give, respectively, the local phonon spectra on the oxygen atom, on one of its nickel neighbours and on a free nickel (111) surface. The arrows indicate the positions of the experimental electron loss peaks [29]. The full lines are for vibrations perpendicular to the surface and the dotted lines for vibrations in the surface plane.

The spectral densities for vibrations of O on Ni(111) surface were also calculated recently by BLACK [30] from a classical force constant model. The positions of the so obtained peaks agree also with the data obtained by IBACH et al [22].

4.6 Conclusion

The physics of adsorbate induced optical phonons seems to be now well understood. With the increasing number of experimental results, one is faced now to propose dynamical models which can fit experimental data. The situation is similar to that of the bulk lattice dynamics when only elastic constants were measured, before the neutron scattering determinations of the phonon dispersion curves. However here we have still often a major unknown: the surface cristallography.

We believe however that thanks to many different surface experiments the surface cristallography will be obtained for more and more systems, as well

as more vibrational frequencies will be measured. Surface lattice dynamics
should begin to mature.

References

1 See for references, A.A. Maradudin, R.F. Wallis and L. Dobrzynski, in
   "Surface Phonons and Polaritons", Handbook of Surfaces and Interfaces,
   3, L. Dobrzynski ed. (Garland STPM Press, New York) 1979.
2 J. Pouliquen, M. Depoorter and A. Defebvre, 4th International Conference
   on Solid Surfaces, Cannes, Sept.1980 and V.R. Velasco, B. Djafar-Rouhani,
   L. Dobrzynski and F. Garcia Moliner, in the same conference, to be published.
3 H. Kaplan, Phys. Rev., 125, 1271 (1962).
4 T. Asahi and J. Hori, in "Lattice Dynamics", ed. R.F. Wallis (Pergamon
   Press, Oxford, 1965) p.571.
5 L. Dobrzynski and D.L. Mills, J. Phys. and Chem. Solids, 30, 1043 (1969).
6 L. Dobrzynski, Surf. Sci. 20, 99 (1970).
7 G. Armand, P. Masri and L. Dobrzynski, J. Vac. Sci. Technol., 9, 705
   (1972).
8 P. Masri and L. Dobrzynski, J. Phys. and Chem. Solids, 34, 847 (1973).
9 G.P. Alldredge, R.E. Allen and F.W. de Wette, Phys. Rev., B 4, 1682
   (1971).
10 W.R. Lawrence and R.E. Allen, Phys. Rev., B 14, 2910 (1976) and B 15,
   5081 (1977).
11 D. Castiel, L. Dobrzynski and D. Spanjaard, Surf. Sci. 60, 269 (1976).
12 L. Dobrzynski and D.L. Mills, Phys. Rev., B 7, 1322 (1973).
13 G. Armand and J.B. Theeten, Phys. Rev., B 9, 3969 (1974); G. Armand and
   Y. Lejay, Solid State Comm., 24, 321 (1977).
14 G. Allan and J. Lopez, Surface Science, to be published.
15 D.W. Bullett and M.L. Cohen, J. of Phys. C, 10, 2083 (1977).
16 B.K. Agrawal, B. Djafari-Rouhani and L. Dobrzynski, Surf. Sci., 89, 446
   (1979) and B.K. Agrawal, Phys. Rev., to be published.
17 N.V. Richardson and A.M. Bradshaw in: Proc. Intern. Conf. on Vibrations
   in Adsorbed Layers, KFA Jülich, 1978, eds. H. Ibach and S. Lehwald, p.2.
18 J.P. Coulomb and P. Masri, Solid State Commun., 15, 1623 (1974).
19 N.S. Gillis, C.R. Fuselier and J.C. Raich in: Lattice Dynamics International Conference Paris, M. Balkanski ed. (Flammarion Sciences, Paris,
   1978) p.341.
20 B. Djafari-Rouhani and L. Dobrzynski in: Proc. Intern. Conf. on Vibrations
   in Adsorbed Layers, KFA Jülich, 1978, eds. H. Ibach and S. Lehwald, p.12.
21 P. Masri, B. Djafari-Rouhani and L. Dobrzynski in: Proceedings of the
   International Conference on Lattice Dynamics, M. Balkanski ed. (Flammarion
   Sciences, 1977) p.312.
22 H. Ibach, M. Bruchmann, Phys. Rev. Lett., 44, 36 (1980).
23 U.A. Jayasooriya, M.A. Chesters, M.W. Howard, S.F.A. Kettle, D.B. Powell
   and N. Sheppard, Surf. Sci., 93, 526 (1980)
24 H. Froitzheim, H. Ibach and S. Lehwald, Phys. Rev. Lett., 36, 1549 (1976).
25 A. Adnot and J.D. Carette, Phys. Rev. Lett., 39, 209 (1977).
26 W. Ho, R.F. Willis and E.W. Plummer, Phys. Rev. Lett., 40, 1463 (1978).
27 M.R. Barnes and R.F. Willis, Phys. Rev. Lett., 41, 1729 (1978).
28 R.F. Willis, Surface Science, 89, 457 (1979).
29 V.R. Velasco, F. Yndurain, Surface Sci., 85, 107 (1979).
30 J.E. Black, to be published.

5. Inelastic Electron Tunnelling Spectroscopy

D. G. Walmsley

With 19 Figures

5.1 Background

Inelastic electron tunnelling spectroscopy (IETS) is fast
becoming established as a useful technique for the study of
vibrational modes of surface adsorbates. It owes its
existence to some observations in the middle 1960's at the
Ford Motor Company Laboratories by R. C. JAKLEVIC and
J. LAMBE [1,2]. They noted anomalies in the current-voltage
characteristic of a metal-insulator-metal thin film tunnel
sandwich. These anomalies showed up more clearly in the first
and particularly the second derivative of the characteristic

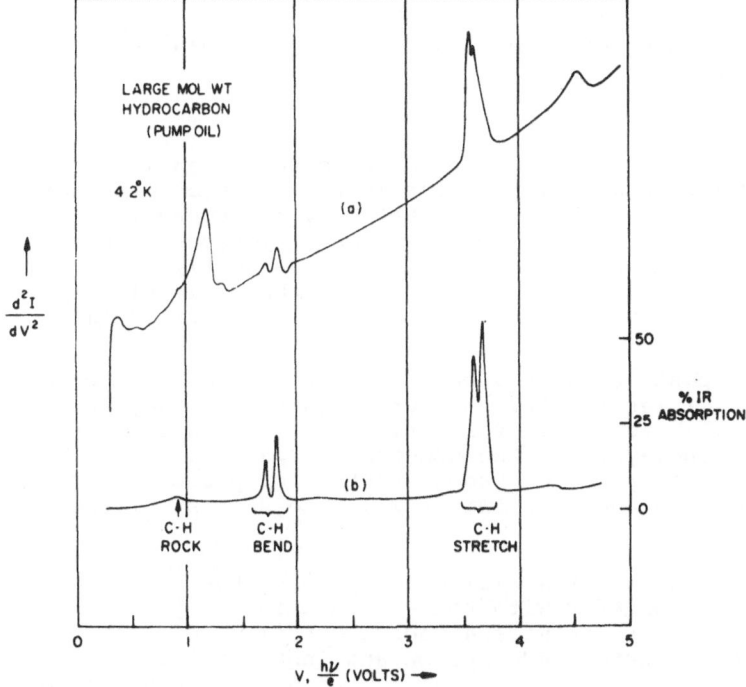

Fig.1 Tunnelling spectrum (a) from large molecular weight
hydrocarbon and infrared spectrum (b) of same material.

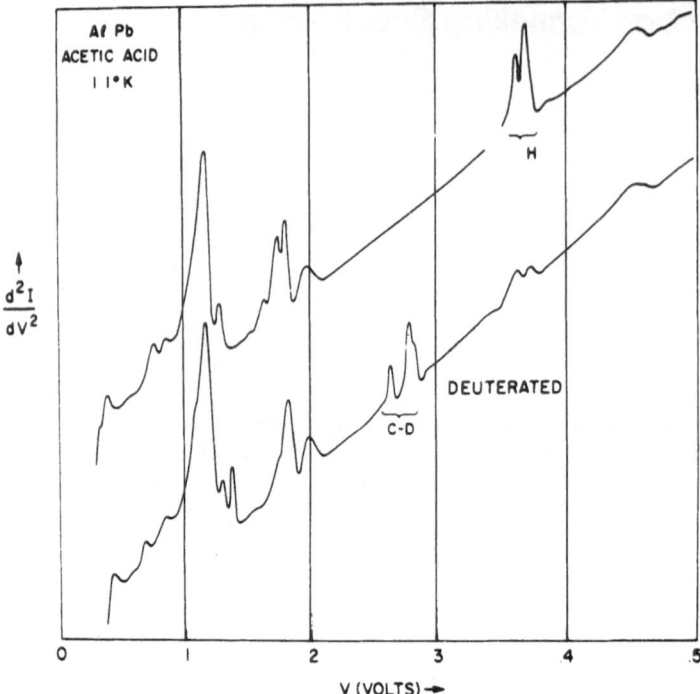

Fig.2 Tunnelling spectra from adsorbed acetic acid and deuterated acetic acid

and were identified as being due to the onset of an inelastic process in which organic impurities in the sandwich were being stimulated to higher vibrational states by the tunnelling electrons. The impurity in question was a hydrocarbon pump oil (Fig.1) and the interpretation was subsequently confirmed through deliberate doping with other molecules and correlating the tunnelling results with known infrared data (Fig.2).

Over the past 10 years a number of other laboratories have developed facilities for IETS measurements until today we have a controlled technique for obtaining good quality spectra of adsorbate monolayers in the spectral range from 300 cm^{-1} to 4000 cm^{-1}. Figure 3 is such a spectrum of phenol adsorbed on aluminium oxide and illustrates well the success of the method.

In this paper the aim will be to present the basics of IETS, compare it briefly with related techniques, give a few examples of where it has made a useful contribution to our understanding of surface adsorption and indicate the directions in which it may develop in the near future. More substantial and comprehensive coverage of the subject is to be found in a recent monograph in the Springer Series in Solid State Sciences [3] and in a number of review papers [4,5,6].

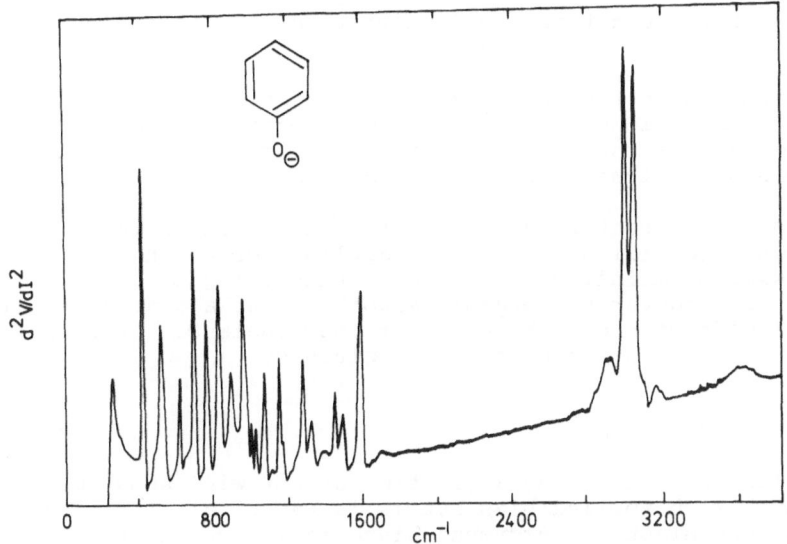

Fig.3 Tunnelling spectrum from phenol adsorbed on aluminium
oxide

5.2 IETS

5.2.1 Principle of IETS

The principle of IETS may be understood by reference to Fig.4
which schematically represents a tunnelling sandwich. The
bottom metal layer is a thin (~1000Å) aluminium film deposited
on a glass microscope slide substrate. The deposition takes
place in a chamber with base pressure in the $10^{-5}-10^{-6}$ Torr
range. Next an oxide barrier is grown on the aluminium to a
thickness of 10-20Å. This may be achieved simply by exposure
of the freshly deposited film to atmosphere, or by admitting
moist oxygen to the film preparation chamber; a particularly
convenient and controllable procedure is to establish an
oxygen dc glow discharge in the chamber with the aluminium
film in the ionised positive column of the discharge. The third
step is to introduce the organic species of interest to the
oxide surface where, hopefully, it will chemisorb. Either
it may be evaporated from a reservoir into the film deposition
chamber or it may be externally applied as a solution to the
oxide surface and excess can then be spun off. Finally the
top metal film of the sandwich is deposited at $10^{-5}-10^{-6}$ Torr.

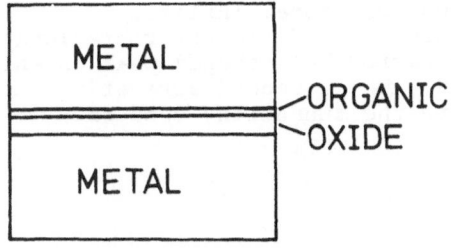

Fig. 4 Schematic
diagram of a doped
tunnel junction

For reasons that are not too well understood lead is the most satisfactory top metal electrode.

The sandwich is removed from its preparation chamber, it is mounted on a long (~1m) probe, two electrical leads are attached to each of the top and bottom metal films, and then it is immersed in liquid helium for measurements.

The tunnel current which flows from one metal to the other when a potential difference is applied across the structure will be mainly due to elastic tunnelling of electrons. If however the organic adsorbate has a characteristic vibrational mode of energy $\hbar\omega$ an additional inelastic process can occur above a minimum bias, V_{min}, given by

$$eV_{min} = \hbar\omega \quad . \tag{1}$$

Like the elastic process the inelastic current will show roughly linear growth with applied bias but it is rather difficult to see against the elastic background since it is typically responsible for a change of only 1 part in 10^3 in the overall conductance of the sandwich. The first or second derivatives of the current-voltage characteristic (Fig.5) are therefore studied.

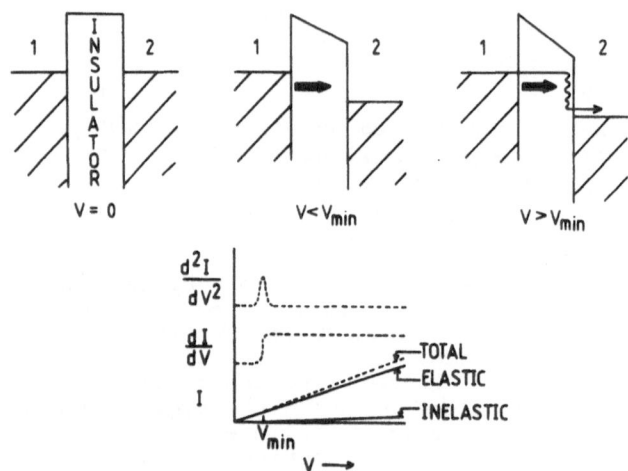

Fig.5 Elastic and inelastic tunnelling

The second derivative displays each additional inelastic threshold, corresponding to the onset of a different vibrational excitation of the adsorbate, as a narrow bell-shaped peak on an otherwise fairly smooth background. These second derivatives d^2I/dV^2, or more often d^2V/dI^2, are the tunnelling spectra.

5.2.2 Nature of Interaction

It is clear from Fig.3 that IETS can give good resolution and sensitivity over nearly the full vibrational spectrum. We should however ask a few questions about its theoretical limitations. Before that we must enquire what is the nature of the interaction between the tunnelling electrons and the organic adsorbate? The accepted picture [2] is that the HARRISON [7] matrix element applies to the tunnelling process. It may be written

$$|M_{12}| \propto \exp\left\{ - \int_0^d dz \left(\frac{2m}{\hbar^2}\right)^{\frac{1}{2}} [U(z) + U_{int}(z) - (E-E_\perp)]^{\frac{1}{2}} \right\} \qquad (2)$$

for an electron of total energy E and energy $E_\perp$ associated with motion parallel to the barrier; the barrier profile is given by $U(z)$ where z is the distance from one face and $U_{int}(z)$ represents the interaction between an electron and the dipole moment of the adsorbate. In the case of species with a permanent dipole moment having a component p_z perpendicular to the barrier the interaction has the form

$$U_{int}(z) = \frac{2ep_z z}{(z^2+r_\perp^2)^{\frac{3}{2}}} \qquad . \qquad (3)$$

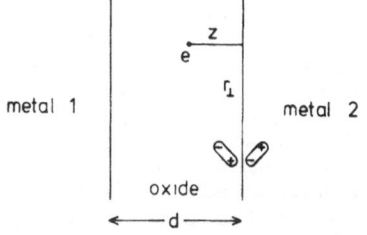

metal 1

metal 2

Fig. 6 Schematic diagram to show geometrical parameters in theory of electron-dipole interaction

Consideration of the image dipole in the top electrode restricts the interaction to the z-component of the dipole and introduces the factor 2; the distances z and $r_\perp$ are shown in Fig.6. If the adsorbate has no permanent dipole moment, a polarisation may be induced by the tunnelling electron and the corresponding interaction term is

$$U_{int}(z) = - \frac{4e^2\alpha z^2}{(z^2+r_\perp^2)^3} \qquad . \qquad (4)$$

With these interactions the predicted spectral intensity of a permanent dipole mode is

$$d^2I/dV^2 \propto |<m|p_z|0>|^2 \qquad (5)$$

which is one component of the matrix element that arises in
infrared spectroscopy, and the predicted intensity of an induced
dipole mode is

$$d^2I/dV^2 \propto |<m|\alpha_{zz}|0>|^2 \qquad (6)$$

where α is an element of the polarizability tensor arising in
Raman spectroscopy. Both intensities should in principle have
a high (viz. $\cos^2\theta$) dependence on the dipole orientation with
respect to the tunnelling direction. It should therefore be
possible to determine adsorbate orientations from a study of
the spectral line intensities. When reasonable physical data is
substituted in the expression for the intensities, conductance
steps of 0.1% can be predicted [8,2] in agreement with
observation.

5.2.3 Linewidths

Two effects cause spectral line broadening. One is the thermal
smearing of the electron distributions in the metal electrodes
at the Fermi energy [2]. At 0 K the inelastic process should
have a sharp onset with a step discontinuity in the tunnelling
conductance and a delta-function peak in the second derivative,
d^2I/dV^2. At non-zero temperatures transitions from partially
filled states above the Fermi energy in one electrode to
partially empty states below the Fermi energy in the second
electrode allow the process to set in at slightly lower bias
and the result is that the sharpness in the onset of the
elastic process becomes smeared. Fig.7 shows the thermal
smearing function. It has a width of 5.4 kT at half height;
thus measurements must be made in the helium temperature
range. At 4.2 K the smearing is 2 meV or 16 cm^{-1} which is
rather greater than the natural linewidth of some of the
vibrational modes being observed. The convenience of
immersing samples in storage dewars has attracted many workers
to do the measurements at 4.2 K. If however the temperature is
reduced to 2 K or 1 K the thermal smearing becomes comparable
with the spectral widths of the sharpest lines and useful
improvement in resolution results. A second advantage of
working below the helium lambda transition is the much improved
heat transfer from the sandwich to the bath. Sometimes large

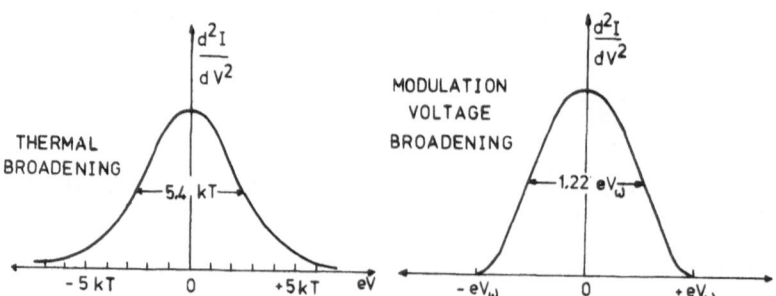

Fig.7 Thermal smearing and modulation smearing functions

heat dissipation in the sandwich during measurement can cause
serious temperature increase and loss of resolution.

The other linebroadening mechanism arises from the method
of measurement. In Fig.8 is shown the circuit we use; other
research groups use the same general principle. A dc current
ramp sweeps the bias across the sandwich from 0 to 0.5V in about
40 mins and the bias is displayed along the x-axis of a
recorder. An ac modulation current at 50 kHz is applied in
parallel and the non-linear characteristic of the sandwich
produces a signal at 100 kHz in proportion to its second
derivative, d^2V/dI^2. This is detected by the PSD (Brookdeal
Model 9503) and displayed along the recorder y-axis. We find
that tuned circuits are particularly helpful in giving good
signal-to-noise performance. We achieve best results at high
audiofrequencies and have a tuning capacitor on the output
transformer of the detection circuit to cope with variations in
sample capacitance from one sample to another. The modulation
is responsible for spectral line broadening. Fig.7 shows the
form of the broadening function [9]. At half height it
contributes $1.22 \ eV_\omega$ where V_ω is the amplitude of the modulation
signal. Clearly the best strategy is to use a modulation signal
which matches the thermal smearing; a larger modulation will
reduce the resolution while a smaller modulation will result in
a lower detected second derivative signal. The latter falls off
roughly as the square of the modulation amplitude.

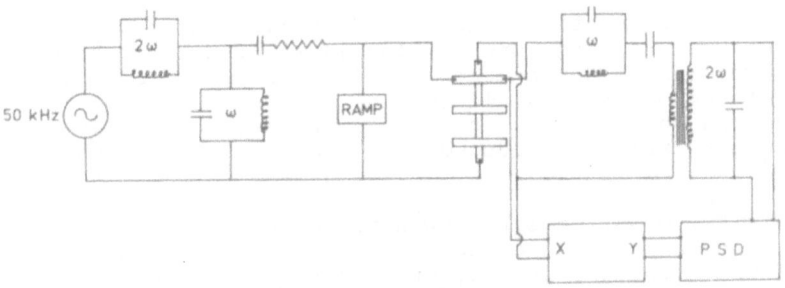

Fig.8 Circuit used for measurement of second derivative, d^2V/dI^2

Even at 1 K there is still significant thermal (and usually
modulation) broadening. It amounts to 0.5 meV or 4 cm^{-1}
and the observed widths of the sharper lines are around 2 meV
(see Fig.3). In order to check out just how great is the
natural width of these lines we have cooled some junctions in
a helium dilution refrigerator at the Rutherford Laboratory to
approximately 0.1 K and examined the spectral lines carefully
[10]. Figure 9a shows the line in the phenolate ion spectrum
at 410 cm^{-1}; the continuous noisy curve is the recorder trace,
taken with the junction in a magnetic field sufficiently strong
to destroy the superconductivity of the metal electrodes. (The
sandwich was of the form Al-Al oxide-Phenol-Pb.) If it is
assumed that the line has zero natural width, and the thermal
and modulation smearing are taken into account, the expected

lineshape is as indicated by the dashed line. Clearly the
observed width is greater and a good fit is obtained (open
circles) to the measured shape if a Gaussian profile with a
natural width of ~1.5 meV or 12 cm⁻¹ is attributed to the line.
In a reduced magnetic field the Pb electrode becomes superconducting with a resultant change in the electron density of
states to the BCS form. If this is included in the calculation
a good fit is obtained to the observation as shown in Fig.9b.

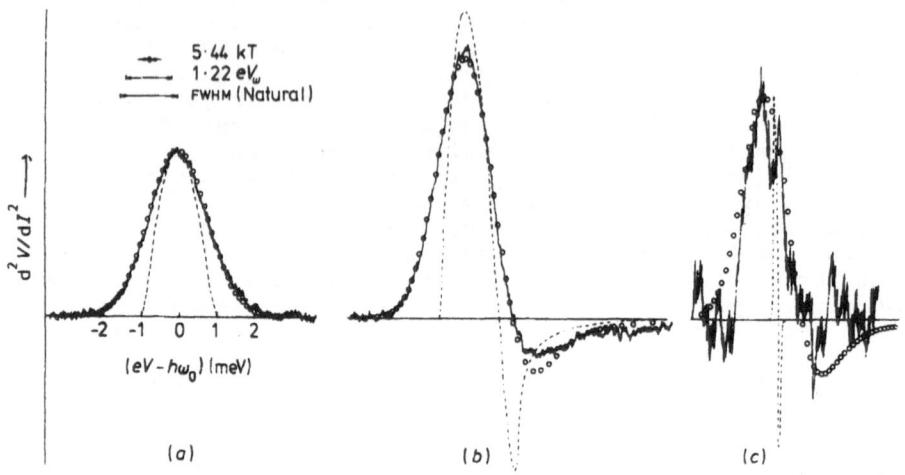

Fig.9 Tunnelling spectral line at 410 cm⁻¹. Continuous
line is recorder tracing, broken line is theoretical prediction
for sharp spectral line and open circles are calculated for
line with 1.504 meV natural linewidth. In (a) both electrodes
are normal, in (b) the Pb electrode is superconducting and
in (c) the modulation is reduced from 1 mV to 0.1 mV [10]

Not only is the observed line now narrower but it also shows
an undershoot on the high energy side as Lambe and Jaklevic
predicted [2]. This latter effect is not readily seen at 1 K
or above. In Figs.9a and 9b modulation smearing dominates
over thermal smearing. In Fig.9c the modulation has been
reduced by a factor of 10 to 0.1 meV. The noise level is now
rather high but the signal fits tolerably well to the
calculated lineshape and confirms the natural linewidth at
1.5 meV. Under these conditions of measurement the resolution
available is ~0.1 meV (0.8cm⁻¹). We have examined other lines
in the spectra of phenol and benzaldehyde adsorbed on aluminium
oxide at 0.1 K and have found no evidence for natural widths
less than 1 meV. It would seem likely therefore that IETS has
the necessary resolution to see all vibrational modes in most
molecular adsorbates. Indeed, as a result of these studies
it is fair to conclude that IETS measurements may be made at
temperatures in the range 1-2 K without significant loss of
information and probably in most cases 4 K will be satisfactory.

5.2.4 Sensitivity and Surface Coverage

Most IET spectra are taken from samples with a monolayer of adsorbate on an area of ~1 mm. The technique is thus very sensitive. If the coverage were reduced to 1% of a monolayer we might still expect to see most of the features present in Fig.3. LANGAN and HANSMA have compared the IET signal from one line, at 686 cm^{-1}, in the spectrum of the adsorbed benzoate ion with its surface concentration (Fig.10). The latter they determined by radioactive labelling and counting [11]. Their conclusion is that the IET signal decreases more rapidly than surface concentration but they can still see signals down to 3% of a monolayer. Also they were able to quantify the maximum surface concentration, or monolayer coverage, at one adsorbed ion per 15$\overset{2}{\text{Å}}$.

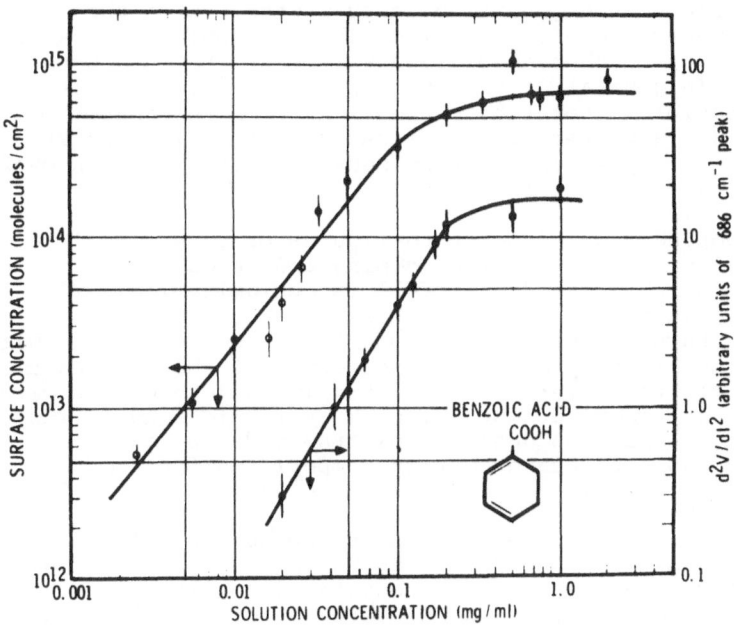

Fig.10 Surface concentration and 686 cm^{-1} line intensity versus concentration of doping solution of benzoic acid [11]

Samples prepared by vapour doping will normally have a full monolayer of adsorbate unless the exposure is carefully controlled. Usually we open the preparation chamber to a reservoir containing the dopant for ~10s. If necessary the reservoir may be heated to increase the vapour pressure of the dopant and occasionally it is helpful to pass a stream of warm nitrogen through the reservoir and preparation chamber at the same time. A quartz crystal microbalance shows the accumulation of dope on the oxide of the sandwich. In most cases several molecular layers are formed but the outer layers desorb again when the chamber is evacuated prior to top electrode deposition. A reproducible amount remains behind characteristic of the particular dopant species. These observations are the basis for arguing that a monolayer is being observed. By and large the IET spectra from a given adsorbate have reproducible intensity as judged against the background elastic signal. Those species such as acids which adsorb readily give strong signals and the spectrometer gain can be kept low. Another useful indication of coverage is the signal-to-noise ratio though this is less reliable. The background noise can be due to the usual factors such as environmental pick-up in the detection circuit or instrument noise. A more variable contribution comes from the samples themselves. When examined carefully at very low temperatures, ~0.1 K, it is clear that samples which might simply be regarded as noisy do have spurious though reproducible additional peaks which contribute to the generally noisy appearance of the background. The cause obviously lies in the sample fabrication and may be due to further chemical reaction of the adsorbate on the surface. In general it is a good policy to cool samples as soon as possible after fabrication. At room temperature they rapidly (in a matter of hours) become more noisy but at liquid nitrogen temperature show no sensible deterioration over periods of weeks.

Liquid doping with either pure liquids or solutions is a widely used alternative to vapour doping and leads to comparable results. The main pitfall is contamination. This may occur through exposure of the oxide to a contaminated atmosphere or through reaction of impurities in the solvent with the oxide. The latter problem becomes acute if the adsorbate is not particularly reactive; it is relatively easy to end up with a spectrum of a reactive impurity such as formic acid coming through stronger than the desired spectrum of an unreactive species which has been beaten out in competition for the reactive surface sites.

Even in samples prepared by vapour doping there is difficulty in seeing a spectrum of an unreactive species. We have frequently observed, with the quartz microbalance, such molecules condense or physisorb on the oxide of a sandwich and remain during subsequent pump-down. Although the high resistance of the completed sandwich showed clearly that dope had been added successfully, no identifiable spectrum could be seen. We are inclined to conclude that if a chemically-bound two-dimensional monolayer is necessary for the technique to work then the nature of the interaction discussed in 2.2 above may be incorrect. Both infrared and Raman spectroscopies are entirely successful in the study of unbound random aggregates. The interaction involved in tunnelling may be rather different; it may be more closely analogous to that in electron scattering from gaseous molecules.

5.3 Oxides

5.3.1 Clean Aluminium Oxide

The vast majority of IETS studies have been concerned with
various species adsorbed on aluminium oxide. It is therefore
important to see what background spectrum is obtained from an
undoped system. Fig.11 shows the result from an Al-Al oxide-Pb
tunnel junction. The structure is closely similar from one
sample to another. Its interpretation has been considered by
several workers. The largest feature, an asymmetric broad band
peaked at ~940 cm⁻¹, is identified as an Al-O mode. It agrees in
position and shape remarkably well with the infrared reflectance
spectrum obtained by MAELAND et al [12] but not with the
transmission spectrum (See Fig.11, first insert). In fact these
authors showed this absorption peak is present only for the
p-polarised reflectance spectrum when the electric field of
the radiation is perpendicular to the metal surface (See Fig.11,
second insert). This situation is analogous to the tunnelling
experiment and in both it is expected that dipoles perpendicular
to the surface will be excited. MAELAND et al studied air
oxidised and anodised aluminium and obtained the same result from
both. Tunnelling studies of air-oxidised and plasma-oxidised
films again show an identical peak in the spectrum. Thus it seems
that the very thin, ~15Å, oxide used in tunnelling studies has much
the same composition and lattice dynamics as the thicker
anodised layers. It does not follow that its surface reactivity
will be the same; that can only be determined by chemisorption
studies.

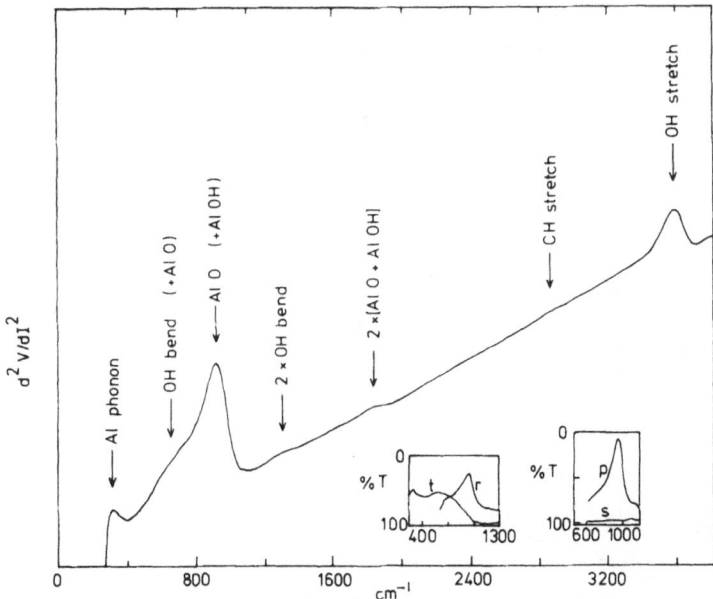

Fig.11 Tunnelling spectrum of undoped Al-Al oxide-Pb junction.
Inset are infrared spectra of the 940 cm⁻¹ region for
comparison [12]

The second obvious peak in the tunnelling spectrum is an OH stretch mode at ~3600 cm^{-1} with a half-height width of ~130 cm^{-1}. It shows the expected $1/\sqrt{2}$ isotope shift when deuterated [2], is always present in clean Al-Al oxide-Pb junctions with about the same intensity relative to the Al-O mode, and in doped junctions is a helpful monitor of surface reactions of adsorbates, as will be discussed below. The presence of OH mode is not surprising in view of the well-established requirement to have slightly moist conditions when oxidising aluminium. It would appear that what grows is a well-defined mixed oxide and hydroxide. The OH bend mode appears at ~600 cm^{-1} and its overtone at 1200 cm^{-1}. There is evidence for some OH, perhaps a bulk hydroxide, bending mode in the ~940 cm^{-1} region with an overtone at ~1880 cm^{-1}. This question has been discussed in detail by BOWSER and WEINBERG [13]. The small peak at 300 cm^{-1} is an aluminium metal lattice mode. Finally, in Fig.11 the small peak at ~2900 cm^{-1} is a CH stretch mode owing to accidental contamination; this is negligible in comparison with the signals observed at this energy in doped junctions. Below 300 cm^{-1} structure due to the metal electrode phonons is dominant.

5.3.2 Formic Acid on Aluminium Oxide

A wide range of species adsorbed on aluminium oxide has by now been studied by IETS. Rather than recount these in review fashion

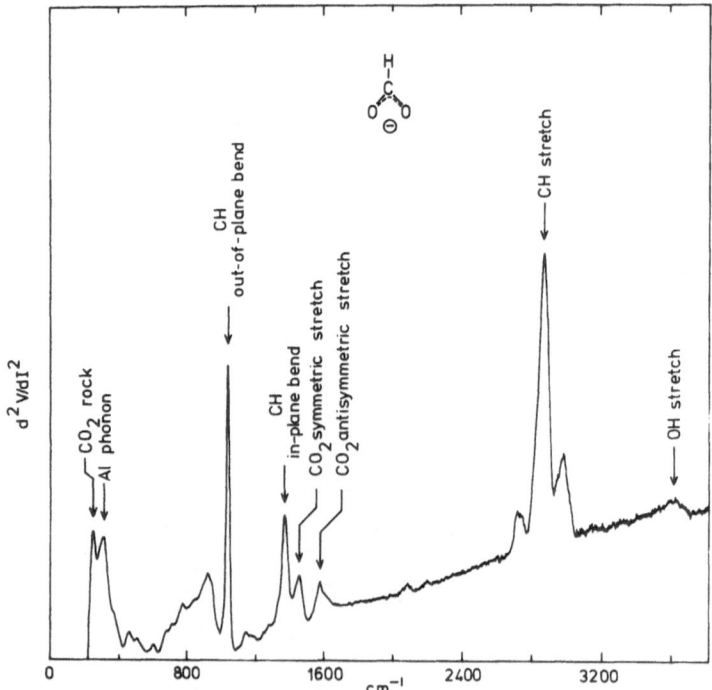

Fig.12 Tunnelling spectrum of formic acid adsorbed on plasma-grown aluminium oxide [14]

it is probably more useful here to consider one or two examples and show what can be learnt by IFTS. Fig.12 shows the tunnel spectrum obtained when formic acid is adsorbed on plasma-grown aluminium oxide [14]. The oxide mode at 940 cm^{-1} is still evident as is the OH stretch mode at ~3600 cm^{-1} though the latter is somewhat reduced in intensity as compared with the undoped sample (Fig.11). If formic acid were physisorbed without reaction we would expect an increase in the OH mode strength at 3600 cm^{-1} or the appearance of an additional mode nearby. Neither is observed. Also there is no evidence for the C=O stretch mode of formic acid which would be expected at ~1780 cm^{-1} or the C-O-H stretch mode at ~1100 cm^{-1}. Instead we may understand the spectrum by postulating the reaction of formic acid with the oxide surface to produce an adsorbed formate ion. The proton liberated in the reaction is not seen as an additional OH mode so we conclude that it combines with an OH species already on the surface to form water which is then pumped away. This accounts for the reduction of the OH peak at ~3600 cm^{-1} and is consistent with the absence of H_2O modes which would be expected at 1630 cm^{-1} and 3300 cm^{-1}. The formate ion with 4 atoms is expected to show 6 distinct normal modes of vibration. The strong CH stretch mode stands out clearly at 2875 cm^{-1}. We expect also to see two CH bend modes one in the plane of the ion and one out-of-plane at lower frequency. Likwise the CO_2^- moiety will have two stretch modes; in one the CO vibrations will be in phase and symmetric with a net oscillating dipole component perpendicular to the oxide surface while in the other the vibrations will be out of phase and the net oscillating dipole will be parallel to the oxide surface. Identification of these modes in the tunnel spectrum requires some care. The most straightforward are the antisymmetric CO_2^- mode at 1580 cm^{-1} which is commonly seen in the infrared spectrum of the formate ion [15,16,17] and the out-of-plane CH bend at 1038 cm^{-1}. The assignment of the modes at 1370 cm^{-1} and 1456 cm^{-1} is less easy. The symmetric CO_2^- mode is seen by IETS at 1463 cm^{-1} in adsorbed acetate ion and at 1444 cm^{-1} in adsorbed propionate ion on plasma grown aluminium oxide. On this evidence there is a fair case for identifying the 1456 cm^{-1} formate mode as the symmetric CO_2^- stretch vibration. Then the 1370 cm^{-1} line corresponds to the in-plane CH bend mode. MAAS [18] has recently summarised previous infrared work on metal formates and highlights the difficulty of making assignments in this region of the formate ion spectrum. Two authors who studied monocrystals with polarised radiation [19,20] did conclude that the CH bend was lower in frequency than the symmetric CO_2^- stretch mode; other authors choose the reverse assignment. In general the IR spectra show up the ~1370 cm^{-1} mode clear and narrow. There is always a broad mode from 1450 cm^{-1} to ~1700 cm^{-1} which clearly includes the antisymmetric CO_2^- mode. This breath is not set by instrument resolution, which can be as good as 3 cm^{-1} or even 1 cm^{-1}, but is a real linewidth. The reason is probably associated with thermal agitation of the CO_2^- moiety and its consequential sampling of different environments; CO_2^- modes are well known for their broad structure in infrared spectra. The infrared spectra do not show a clear mode at 1456 cm^{-1} which we see well in IETS and wish to assign as the symmetric CO_2 mode. The low temperature of the IETS measurement may account for the sharpness of this mode; bulk measurements on formates at 4.2 K would be interesting. (JONES and McLAREN [21] found better resolution of acetate infrared spectra at 80 K.)

The last line requiring assignment is the CO_2^- rocking mode which appears at 242 cm^{-1}. The small peaks on either side of the CH stretch mode are assigned to an overtone of the CH in-plane bend mode and a combination mode of the symmetric and antisymmetric CO_2^- modes. Table 1 summarises the information.

Deuterated formate ion on plasma-grown aluminium oxide gives support to these assignments [14].[1] The C-D stretch is at 2144 cm^{-1} and the C-D out-of-plane bend at 905 cm^{-1}. The two CO_2^- modes persist with much the same intensity as before but the mode at ~1370 cm^{-1} is considerably reduced having a broad low energy tail which would support the idea of the 1370 cm^{-1} formate mode being a CH bend and the tail in the spectrum of the deuterated species could be the corresponding CD bend.

A striking feature of the IET spectrum of adsorbed formate and a feature that recurs in other IET spectra is the strength of the CH modes. These are much stronger than in infrared or Raman spectra.[2] Of course most organic adsorbates will have many CH bonds and the information on them is indeed welcome. The resolution of IETS is such that one can clearly see the distinction in frequency between CH stretch mode of an aromatic ring compound and that associated with a saturated hydrocarbon chain or that of a doubly or triply bonded carbon atom. The strength of the CH modes is something of a puzzle. It casts doubt once again on whether the interaction between the tunnelling electron and adsorbate is as described in 2.2. Gas phase electron diffraction shares this high sensitivity to CH. Perhaps billiard ball kinetics is more appropriate to the IET problem; in such a treatment protons will show up to relatively good advantage. (If only 0.1% of the tunnelling electrons induce inelastic interactions, it may be because only those passing very close to the adsorbed ions are effective.) In such circumstances a far-field plane electromagnetic wave treatment, appropriate to infrared or Raman spectroscopy, may be totally invalid. Selection rules will disappear and information on adsorbate orientation will not be easily deduced if our suspicions are borne out.

5.3.3 Clean Magnesium Oxide

A spectroscopy limited to studies of adsorbates on only one oxide is not likely to become important. Hence there is an interest in extending the method to other systems. The success and convenience of aluminium oxide is underwritten by two

[1] The assignments of [14] are slightly revised is this discussion.

[2] I am indebted to Prof. Sheppard for emphasising this feature of IETS.

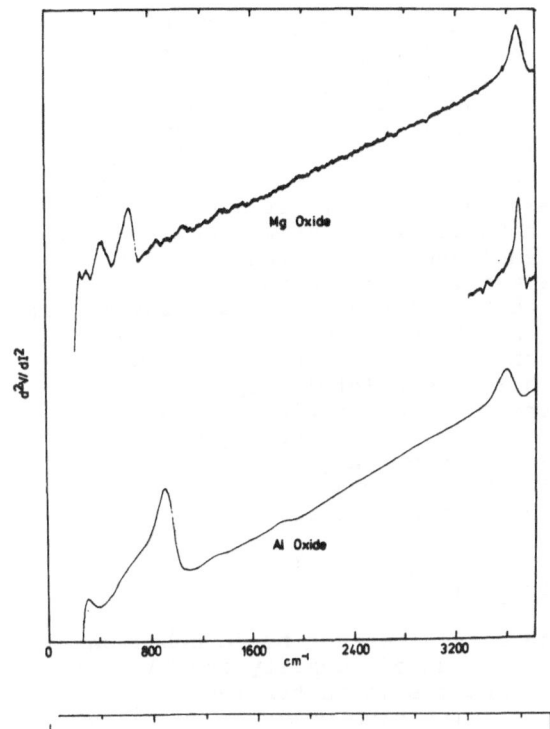

Fig. 13 Tunnelling spectrum at 2 K of undoped Mg-Mg oxide-Pb junction and spectrum at 4.2 K of undoped Al-Al oxide-Pb junction. Inset shows OH stretch mode of Mg junction at 1.1 K

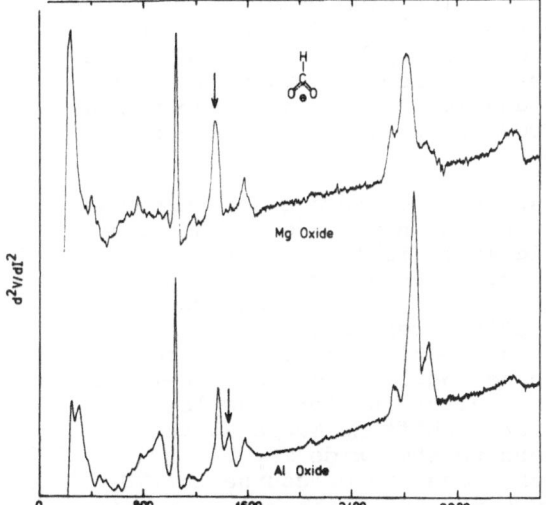

Fig. 14 Tunnelling spectrum at 2 K of formic acid on plasma grown magnesium oxide and on plasma grown aluminium oxide. The arrow denotes the line which has shifted down on magnesium oxide to coincide with another line

properties. It offers a high electrostatic potential barrier (~2V [22,23]) to the tunnelling electrons which in turn allows bias voltages up to 0.5V to be applied without serious growth of the elastic background current and consequent heat damage to the junction as measurements are made towards the higher energy end of the vibrational spectrum. More important still is the presence of hydroxyl ions in the oxide since these are

responsible for much of the interesting chemistry of the
surface; indeed they may be essential to the chemisorption of
most species taking place at all. There are few oxides which
are known to match up to these requirements and which in
addition can be grown on the parent metal as a coherent thin
(~20Å) tunnel layer. An earlier report by KLEIN et al [9]
of the tunnel spectrum of clean and water doped Mg-Mg oxide-
Pb junctions prompted us to examine this particular system.
The clean spectrum is shown in Fig.13 where it may be compared
with that from the aluminium analogue. Two bands at 430 and
645 cm^{-1} are thought to arise mainly from MgO phonon modes
[9]; the introduction of D_2O to the plasma discharge leaves them
substantially unchanged. An OH stretch band is present at
3675 cm^{-1}. It varies quite considerably in intensity from
sample to sample. It can be almost completely replaced by an
OD stretch mode if D_2O is introduced to the plasma discharge
during the oxidation process. So the oxide which grows on
magnesium in an oxygen plasma does contain hydroxyl ions and
also, as may be seen in Fig.13, makes a reasonably constant
background elastic contribution to the spectrum.

5.3.4 Formic Acid on Magnesium Oxide

It turns out that a wide range of species can be adsorbed on
the plasma-grown magnesium oxide and good quality tunnel
spectra obtained [24,25]. The process is rather less
reproducible than in the case of aluminium oxide. One of the
problems is the variation in the amount of OH present in the
oxide. It obviously depends on the vapour pressure of water
in the preparation chamber but also seems to vary with the
rate of evaporation of the magnesium film. High evaporation
rates (50-100 Ås^{-1}) lead to most reproducible low noise
spectra.

Figure 14 compares the spectrum of formic acid adsorbed
on magnesium oxide with that on aluminium oxide. They are
almost identical and the assignments follow as before
(Table 1). One line however has disappeared. It is the
1456 cm^{-1} mode but we see that the 1370 cm^{-1} mode has grown
in height and extends to slightly lower energy. We do not
expect internal modes of the adsorbate to shift as a result of
the change of oxide; only van der Waals forces are involved
and these cause no significant shifts in, for example,
liquids. However we can expect a shift in modes at the site
of chemisorption since the bond to the oxide may be
substantially different. Such modes are of course the CO_2^-
stretching vibrations. Of these the symmetric mode with its
oscillating dipole, having a component perpendicular to the
surface, is most likely to be affected. And the 1456 cm^{-1}
line on aluminium oxide was tentatively assigned to that mode
in 3.2. So the evidence here suggests that we have the
correct assignment and that the mode shifts down to coincide
with or fall slightly below the CH in-plane bend mode. The
decrease in intensity of the combination mode at ~3000 cm^{-1}
lends support to the arguments. We have found closely similar
shifts in the tunnel spectra of acetate and propionate
adsorbates [24,25]. The shifts are interesting in that they

Table 1 IETS vibrational modes of formate anion on plasma-grown magnesium and aluminium oxides

Magnesium oxide host (cm^{-1})	Aluminium oxide host (cm^{-1})	Assignment
236	242	CO_2^- rock
1046	1038	CH out-of-plane bend
1348		$\begin{cases} CO_2^- \text{ symm. stretch} \\ +CH \text{ in-plane bend} \end{cases}$
	1370	CH in-plane bend
	1456	CO_2^- symm. stretch
1575	1580	CO_2^- asymm. stretch
2709	2724	$\begin{cases} CH \text{ in-plane bend} \\ \quad \text{overtone} \end{cases}$
2815	2875	CH stretch
2976	2991	$\begin{cases} CO_2^- \text{ symm. stretch} \\ +CO_2^- \text{ asymm. stretch} \\ \quad \text{combination} \end{cases}$

indicate a stronger electrostatic binding of the carboxylates to surface magnesium cations than to surface aluminium cations. The magnesium oxide lattice is more ionic and it is also a more basic oxide than that of aluminium. Both factors will operate in the correct sense if they are responsible for the shifts.

5.4 Comparison

5.4.1 Comparison with Electron Energy Loss Spectroscopy (EELS)

High resolution electron energy loss spectroscopy is now being applied to the study of organic adsorbates on metals. Although the resolution (~8 meV) is some order of magnitude greater than the narrowest lines in the vibrational spectra, the information it yields is a welcome complement to IETS. In Fig.15 the IET spectrum of the acetate ion on aluminium oxide is shown. Inset is a recent measurement by SEXTON [26] of the EEL spectrum of acetate ion adsorbed on the (100) face of a copper single crystal. Though identical spectra are not expected in view of the different absorbents a comparison is interesting. The arrows identify four observed modes in the EEL spectrum and, as a guide, are reproduced over lines in closely similar positions in the IET spectrum. Clearly the EEL spectrum shows fewer modes. Once more the strength of the CH modes in IETS is highlighted. The CH stretch modes are very weak in the EEL spectrum. If we identify the mode at ~1040 cm^{-1} with a CH_3 rock (as in [14] and not C-C stretch as in [26]) the intensity contrast may be regarded as similar in origin. Our

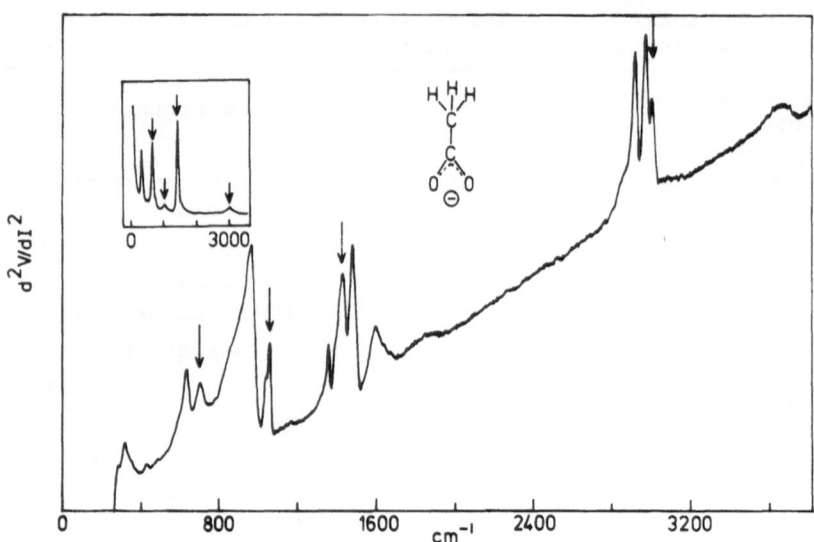

Fig.15 Tunnelling spectrum at 2 K of acetate ion on aluminium oxide. Inset is EEL spectrum of acetate ion on Cu(100) [26]. Arrows are explained in text

Table 2 Comparison of vibrational modes of acetate anion on aluminium oxide observed by IETS with acetate ion on copper observed by EELS

I E T S (cm^{-1})	E E L S (cm^{-1})	Assignment
	339	Cu-O stretch
689	677	CO_2^- symm. def.
948		C-C stretch
1044	1041	CH_3 rock
1411		CH_3 def.
1463	1434	CO_2^- symm. stretch
1583	1602	CO_2^- asymm. stretch
2910⎫		
2964⎬		
3000⎭	3000	CH stretch

recent observations (unpublished) of the acetate ion on magnesium oxide suggest that we should interchange our earlier assignment [14] of the mode at 1411 cm^{-1} to CH_3 deformation and of the mode at 1463 cm^{-1} to symmetric CO_2 stretch. The

EEL spectrum shows the latter more clearly but hardly any signal for the expected asymmetric CO_2 mode. That contrast is quite elegant in its support of arguments concerned with the orientation of the CO_2 moiety on the surface [26]. The IET spectrum shows both modes and the orientation argument is less clear-cut. More than anything this difference probably derives from the contrast between a metal and insulating absorbent surface. Table 2 lists some of the prominent lines in the two spectra for convenience of comparison; the assignments are plausible but not irrevocable.

Electron energy loss spectroscopy has the additional convenience over IETS that is allows the spectrum of an adsorbate to be monitored in real time, which can be a major advantage if temperature-dependent reactions are being studied.

5.4.2 Comparison with Surface Infrared and Raman Spectroscopies

Figure 16 is a comparison first shown by HANSMA of the infrared Raman and IET spectra of benzaldehyde adsorbed on aluminium oxide. It illustrates well the good resolution and high sensitivity particularly to CH modes of tunnelling and is typical of a variety of adsorbates on this oxide. A more detailed comparison is to be found in [4]. In fairness it should be admitted that both infrared and Raman are more versatile in their application to a range of absorbents. There is now the prospect too that surface enhanced Raman spectroscopy [27] (SERS) may prove to be a widespread phenomenon which will enhance the competitive position of that technique.

5.5 Adsorbate Orientation

It is interesting to know not only the chemical nature of an adsorbed monolayer but also its orientation. Here tunnelling is expected to be informative because the accepted theory, outlined in 2.2, predicts intensities varying as $\cos^2\theta$ where θ is the angle that the dipole in the adsorbate makes with the tunnelling direction. However, several factors complicate the issue. For example the lower surface of the top electrode is not planar and so the angle θ is not well defined. Also, all adsorbates are polyatomic and the vibrational modes are those of the molecule or ion as a whole rather than purely deriving from a single bond within the structure. Plausible attempts have been made to make arguments about the orientation of carboxylate [28] and sulphonate [29] moieties in adsorbed acids on aluminium oxide but these depend crucially on correct identification of modes in a rather confused region of the spectrum. One set of modes which is free of this difficulty and which is also relatively independent of coupling with other modes of the adsorbate is the CH stretch modes at the high energy end of the spectrum. Consider the spectrum of p-cresolate shown in Fig.17. There are two broad CH stretch bands. That at ~3000 cm^{-1} is due to ring CH stretching while the lower band at ~2900 cm^{-1} is associated with methyl group vibrations [30]. If the cresolate ion is standing

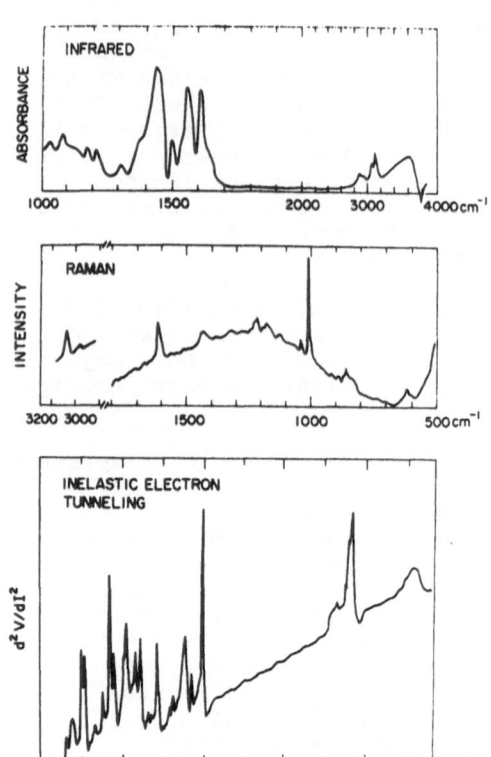

Fig. 16 Comparison of in-
frared, Raman and IET spectra of benzaldehyde adsorbed
on aluminium oxide [4]

Fig. 17 Tunnelling spec-
trum at 2 K of p-cresolate
[30]

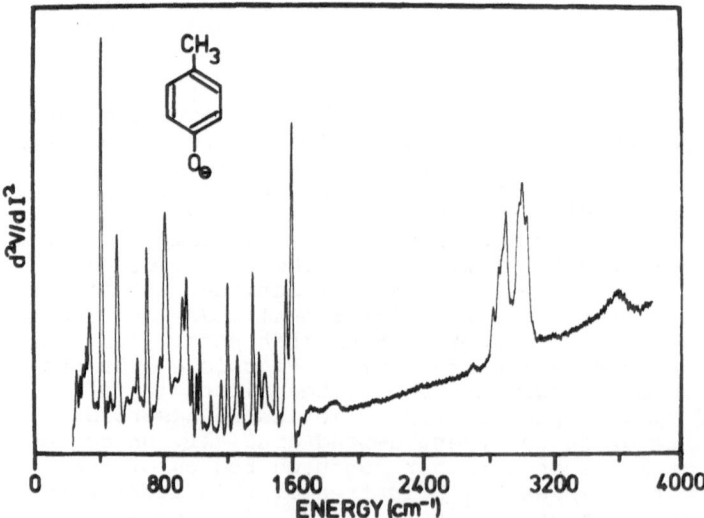

perpendicular to the surface the methyl tripod and ring CH
modes will all be canted away from the direction of
tunnelling and none is expected to show up especially strongly.
That is what is observed. Consider now the spectrum of

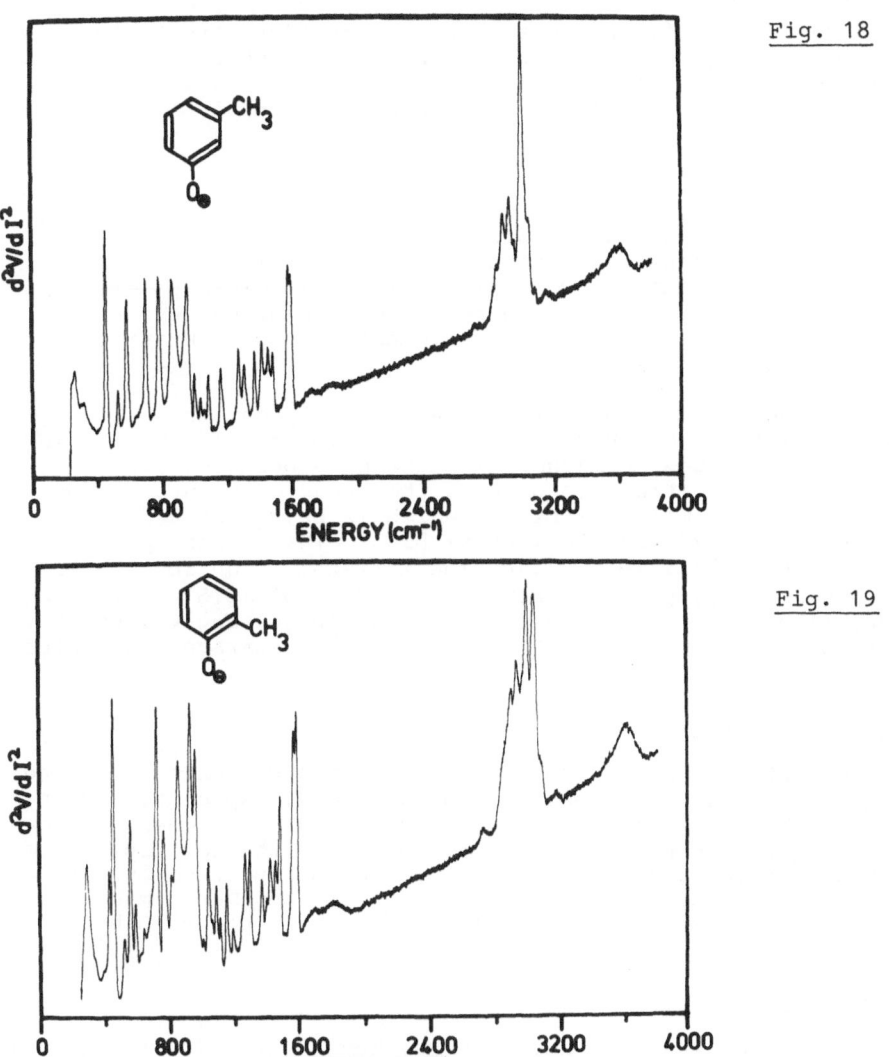

Fig. 18

Fig. 19

Fig.18   Tunnelling spectrum at 2 K of m-cresolate [30]

Fig.19   Tunnelling spectrum at 2K of o-cresolate [30]

m-cresolate shown in Fig.18. The ring CH mode at the para
position is now in the tunnelling direction if the previous
orientation assumption is made and we can expect to see a
strong mode appear in the spectrum. And that is observed.
Finally, if we look at the spectrum of o-cresolate in
Fig.19 we do not see such a strong identifiable ring mode.
The steric demand of the methyl group is likely however to
cant the axis of the molecule away from the tunnelling
direction and so again the CH mode in the para position is not

specially favoured. Whether these observations are fortuitous or can be confirmed in other adsorbates remains to be seen. If they can, the information on adsorbate orientation available from tunnelling may prove highly useful.

5.6 Conclusion

In the limited space available I have tried to give some introduction to the IETS technique. I have illustrated its utility by reference to work with which I am closely familiar. Much other important work has received no mention here and the interested reader should supplement this presentation with the recent reviews [3,4,5,6].

The development of IETS depends heavily on its applicability to other surfaces in addition to aluminium oxide. Already adsorbates on magnesium oxide have been observed [24,31]. An exciting prospect is the study of adsorbates on oxide-supported metal particles which is being pursued by HANSMA et al [32] and KLEIN et al [33]. This opens up a wide range of possibilities. Given the intrinsic high resolution and sensitivity, particularly to CH modes, of IETS there seems every likelihood that it will occupy an increasingly important position in the vibrational spectroscopy of surface adsorbates.

Acknowledgements

I am grateful to Dr. N. M. D. Brown who has patiently educated me in the vibrational spectroscopy of organic species, and to I. W. N. McMorris, W. E. Timms, R. B. Floyd, W. J. Nelson and R. J. Turner who have contributed hugely to the development of IETS in our laboratory. The work has been supported by the Science Research Council, The Northern Ireland Department of Education and Imperial Chemical Industries Corporate Laboratory.

References

1   R. C. Jaklevic, J. Lambe: Phys. Rev. Lett. 17, 1139 (1966)

2   J. Lambe, R. C. Jaklevic: Phys. Rev. 165, 821 (1968)

3   T. Wolfram (ed.): Inelastic Electron Tunnelling Spectroscopy, Springer Series in Solid-State Sciences, Vol. 4 (Springer-Verlag, Berlin, Heidelberg, New York, 1978)

4   P. K. Hansma: Phys. Rept. 30C, 145 (1977)

5   W. H. Weinberg: Ann. Rev. Phys. Chem. 29, 115 (1978)

6   P. K. Hansma, J. Kirtley: Accts. Chem. Res. 11, 440 (1978)

7   W. A. Harrison: Phys. Rev. 123, 85 (1961)

8   D. J. Scalapino, S. M. Marcus: Phys. Rev. Lett. 18, 459 (1967)

9   J. Kelin, A. Leger, M. Belin, D. Defourneau,
M. J. L. Sangster: Phys. Rev. $\underline{B7}$, 2336 (1973)

10   D. G. Walmsley, R. B. Floyd, S. F. J. Read, Journal of
Physics C: Solid State Physics $\underline{11}$, L107 (1978)

11   J. D. Langan, P. K. Hansma: Surface Science $\underline{52}$, 211 (1975)

12   A. J. Maeland, R. Rittenhouse, W. Lahar, P. V. Romano:
Thin Solid Films $\underline{21}$, 67 (1974)

13   W. M. Bowser, W. H. Weinberg: Surface Science $\underline{64}$, 377 (1977)

14   N. M. D. Brown, R. B. Floyd, D. G. Walmsley: Journal of
the Chemical Society, Faraday Transactions II $\underline{75}$, 17
(1979)

15   K. Ito, H. J. Bernstein: Canad. J. Chem. $\underline{34}$, 170 (1956)

16   C. J. H. Schutte and K. Buijs: Spectrochim. Acta $\underline{20}$, 187
(1964)

17   J. D. Donaldson, J. F. Knifton and S. D. Ross: Spectrochim.
Acta $\underline{20}$, 847 (1964)

18   J. P. M. Mass: Spectrochim. Acta $\underline{33A}$, 761 (1977)

19   R. Newman: J. Chem. Phys. $\underline{20}$, 1663 (1952)

20   T. L. Charlton, K. B. Harvey: Canad. J. Chem. $\underline{44}$, 2717
(1966)

21   L. H. Jones, E. McLaren: J. Chem. Phys. $\underline{22}$, 1796 (1954)

22   R. B. Floyd, D. G. Walmsley: Journal of Physics C: Solid
State Physics $\underline{11}$, 4601 (1978)

23   M. F. Muldoon, R. A. Dragoset, R. V. Coleman: Phys. Rev.
$\underline{B20}$, 416 (1979)

24   D. G. Walmsley, W. J. Nelson, N. M. D. Brown, R. B. Floyd:
Applications of Surface Science $\underline{5}$ (June 1980) (to
be published)

25   D. G. Walmsley: unpublished results

26   B. A. Sexton: Chem. Phys. Lett. $\underline{65}$, 469 (1979)

27   J. A. Creighton: these proceedings

28   J. T. Hall, P. K. Hansma: Surface Science $\underline{77}$, 61 (1978)

29   J. T. Hall, P. K. Hansma: Surface Science $\underline{71}$, 1 (1978)

30   I. W. N. McMorris, N. M. D. Brown, D. G. Walmsley: J. Chem.
Phys. $\underline{66}$, 3952 (1977)

31   C. S. Korman, J. C. Lau, A. M. Johnson, R. V. Coleman:
     Phys. Rev. B19, 994 (1979)

32   P. K. Hansma, W. C. Kaska, R. M. Laine: J. Amer. Chem. Soc.
     98, 6064 (1976)

33   J. Klein, A. Leger, S. de Cheveigne, C. Guinet, M. Belin,
     D. Defourneau: Surface Science 82, L288 (1979)

6. Inelastic Molecular Beam Scattering from Surfaces

B. Feuerbacher

With 15 Figures

6.1 Background

An experimental probe for the investigation of surface dynamical properties
would ideally fulfil a number of distinct requirements. First, it should be
capable of a momentum transfer between probe and surface which ranges least
up to the dimensions of the Brillouin zone. This is necessary in order to
allow the investigation of the momentum dependence of the excitations involved, viz. the dispersion relations. Neutron spectroscopy exhibits this
property and has therefore been the most successful tool for the investigation
of the dynamical properties of bulk solids. Application of this technique to
surfaces is limited due to the small cross sections involved. The second requirement is a probe energy in the vicinity of the energies of the excitations
under study, i.e. the phonon energies. This is desirable in order to achieve
high spectroscopic accuracy and to avoid resolution problems as encountered
in techniques based on small differences between large energies, such as
electron energy loss spectroscopy or Raman spectroscopy. Finally, one would
require a high cross section for vibrational excitations *at the surface*,
possibly combined with a small cross section for the corresponding excitations
in the volume in order to avoid ambiguities.

Inelastic scattering of low energy neutral particles is a developing technique that seems to fulfil all of the above criteria. Molecular beams with
high velocity monochromacy can be produced today with energies from a few to
several hundred milli-electron volts. Their momenta fall conveniently into
the range spanned by surface reciprocal lattice vectors. Low energy neutral
particles are not penetrating, so they probe exclusively the surface. At the
energies involved, they form an extremely soft probe that does not destroy
or change the system under investigation.

Molecular beam techniques are by no means novel. They have been used extensively in the realm of gas-phase scattering for the investigation of atomic
and molecular properties. On surfaces, research has been stimulated by the
gas-surface interaction problems encountered with low-orbit satellites.
Studies directed on particular surface problems are now receiving attention
[1,2], such as the investigation of the detailed gas-surface interaction
potential by means of selective absorption techniques [3] or diffractive
scattering, observing structures of clean [4] and adsorbate covered surfaces
[5] . It is mainly the property of extreme surface sensitivity exploited here,
where the experimental data allow a mathematical modelling of the surface [6]
that is unachievable in conventional LEED techniques which suffer from
multiple scattering problems. However, the techniques required for the study
of dynamical surface properties, viz. inelastic Molecular Beam Scattering

(IMBS), are still being developed. This is in spite of the specific advantages outlined above, and therefore illustrates the particular experimental difficulties inherent in the approach. Neutral particles are difficult to detect, and no techniques exist for focussing them or changing their velocity. In addition, the high gas loads unavoidable during the generation of a molecular beam are not only in conflict with the UHV requirements in surface spectroscopies, but they also limit the achievable signal to noise ratios.

At the present time data on IMBS are scarce, but a rapid development of the technique is noticeable over the last few years. The data available to date do indicate that the method is viable for the investigation of surface dynamics. The present paper therefore will be somewhat speculative in nature. It is focussed on the prospects and capabilities of the technique, in particular in the context of the conventional methods presently used to study surface vibrations. The paper will describe the experimental techniques and the inherent problems encountered. Some of the results available will be presented. The main aim of the article remains, however, an attempt to give an introduction to this new experimental approach to a reader who is familiar with the general problems arising in the investigation of vibrational excitations at well defined surfaces.

6.2 Atom Scattering from Surfaces

Scattering of a molecule or atom from a solid surface can be illustrated as shown in Fig.1. The particle is characterized by its mass, energy (or velocity), and the incidence angle Θ_i, so an incident k-vector $\underline{k}_i$ may be defined. The scattered beam is given by two angles, Θ_f and Θ_f, and an energy E_f different from the incident energy E_i for the inelastic case. As the mass remains unchanged, the scattering conditions are described uniquely by the final k-vector $\underline{k}_f$. It is convenient to restrict the geometry to planar scattering, where the scattered beam is observed in the plane formed by the incident beam and the surface normal. Let us first consider the simple case of elastic scattering as shown in Fig.2a. For an elastic event, energy remains unchanged, $E_i = E_f$, so the energy uncertainty ΔE vanishes. We now make use of the Heisenberg uncertainty relation

$$\Delta E \cdot \Delta t \simeq h \quad . \tag{1}$$

As $\Delta E = 0$, we conclude Δt, the time uncertainty, is infinite. This means that an elastic experiment cannot give time-resolved information, such as required for the study of dynamic processes. It provides data relevant

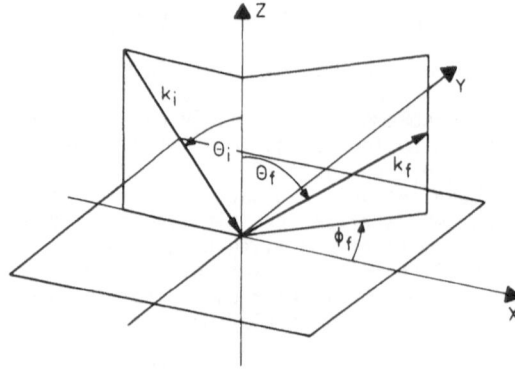

Fig.1 Schematic of inelastic scattering at a surface.

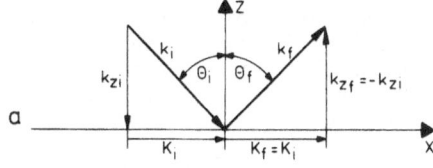

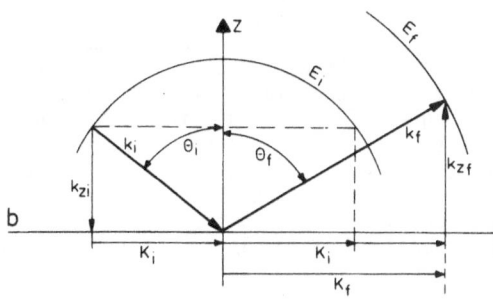

Fig.2 In-plane scattering for the elastic case (a) and the inelastic case (b)

exclusively for either stationary properties (like surface structures) or time averaged properties (like a Debye-Waller factor).

Similar conclusions can be drawn analoguous for the momentum change. In the specular elastic scattering process the momentum components parallel to the surface remain unchanged, $\underline{K}_i = \underline{K}_f$ or $\Delta\underline{K} = 0$ (with the convention $\underline{k} = (\underline{K}, k_z)$). Using now the uncertainty relation between momentum and real space coordinate $\underline{X} = (x,y)$ in the two-dimensional surface space

$$\Delta\underline{K} \cdot \Delta\underline{X} \simeq h \tag{2}$$

we note an infinite uncertainty of the real space coordinate $\Delta\underline{X} = \infty$. Consequently specular scattering will provide information relating to properties delocalized parallel to the surface, and no spatial resolution is possible. These considerations can of course be expanded to the case of a periodic surface structure, where diffraction is possible. It is apparent that diffractive scattering gives information on the large-scale periodicities, but not on the spatial structure within the unit cell.

The situation is quite different as regards the normal component of the k-vector k_z. Specular scattering reverses the sign of k_z, so $\Delta k_z = 2 \cdot k_z$. Therefore the spatial range normal to the surface contributing to the signal is

$$\Delta z \simeq h/2k_z \quad . \tag{3}$$

As the momentum k_z is of the size of the Brillouin zone or larger the spatial uncertainty in the surface normal direction shrinks to the size of a unit cell (layer distance) or less. Any information extracted from the scattering experiment relates to a limited region Δz in real space, which is the quantitative way of expressing the extreme surface specificity of molecular beam scattering.

The same approach can be used for the inelastic case, as illustrated in Fig.2b. Now $\Delta E = E_i - E_f$, so $\Delta t \simeq h/\Delta E$ ensures information on dynamic properties in the region of phonon frequencies for ΔE in the order of a few ten meV. The change in the normal momentum is similar to the elastic case, so the same grade of surface sensitivity is achieved. But now we get a transfer of momentum parallel to the surface, ΔK, which provides spatial information in the surface plane or, in other words, the momentum information required for the measurement of dispersion relations.

As it turns out that the usefulness of a spectroscopic method for dynamical investigations can be judged simply from the amount of momentum and energy transfer, we have compiled these quantities for a number of common techniques in Table 1. Here the momentum transfer ΔK and Δk_z parallel and normal to the surface are expressed in units of reciprocal lattice vectors $G = \frac{\pi}{a}$, where a is the lattice constant. The quantity Δz gives the amount of layers that contribute to the signal and thus the degree of surface specificity of the method. It should be noted here that a technique can well be surface sensitive without being surface specific. Such is the case, e.g., if the signal detected is specific to an atomic species, so it can identify an adsorbed species provided it is different from the bulk species. This is the reason why LEELS can observe adsorbate vibrations from fractions of monolayers, while it has difficulties to detect intrinsic phonons.

Starting with photons in Table 1, we note that no energy resolution can be achieved in x-ray scattering, so only structural information is obtained

Table 1 Comparison of surface vibration spectroscopies

PROBE / TECHNIQUE	E [eV]	ΔE [meV]	$\Delta K_{||}$ π/a	Δk_z π/a	Δz LAYERS
PHOTONS					
X-Ray Scattering	10^4	0	0	0	∞
Raman & Brillouin Scattering	10	10	10^{-4}	10^{-3}	10^3
Infrared Spectroscopy	10^{-2}	10	10^{-4}	10^{-4}	10^4
ELECTRONS					
LEED	10^2	0	0	10^{-1}	10
Energy Loss Spectr.	1-10	100	10^{-2}	10^{-2}	100
NEUTRONS					
Inelastic Scattering	10^{-1}	10	1	0	∞
MOLECULES					
IMBS	10^{-1}	10	1	1	1

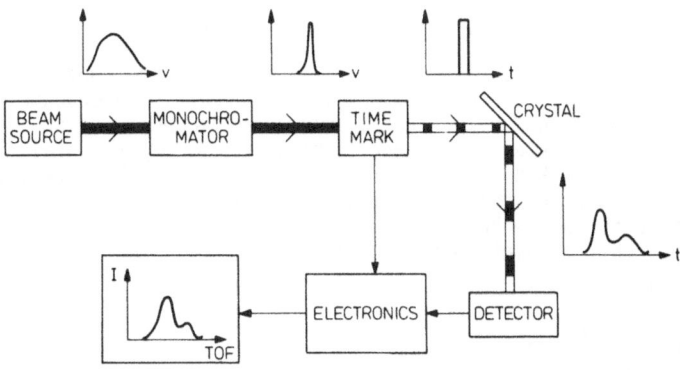

Fig.3 Schematic of the major constituents of an experimental system for measuring inelastic molecular beam scattering.

(this statement does not apply for the EXAFS technique). Inelastic optical probes and infrared spectroscopy do transfer energy in the phonon range, but the intrinsic momentum of photons is so small that it vanishes on the scale of surface momenta. For the same energy transfer, the inelastic techniques transfer a larger normal momentum (due to the larger momentum of the re-flected phonon) and are therefore more surface specific. Electrons are more heavier probes and consequently are more surface specific, but the momenta transferred parallel to the surface plane are still small compared to the Brillouin zone dimensions. Neutrons cover both the requirements of energy and parallel momentum transfer, but, since they are a forward scattering probe, their surface sensitivity is negligible. The neutral molecule, on the other hand, combines the virtues of neutron spectroscopy with the extreme surface specificity desirable for surface dynamics studies.

6.3 Experimental Considerations

The principle of a molecular beam scattering instrument is shown schematically in Fig.3. A beam source is required to produce a high density stream of the chosen atoms or molecules. If the velocity distribution is wide, like for a Knudsen cell source, a speed monochromator is required that usually consists of a set of rotating slotted disks. Other sources, like the free jet source, have a sufficiently narrow velocity distribution so no monochromator is re-quired. Next, a time structure is imposed on the monochromatic beam in order to facilitate time-of-flight detection after scattering. High frequency choppers serve well for this purpose. Chopping before scattering instead of after scattering is only possible if the residence time at the surface is small compared to the flight time. This is, in any case, a prerequisite for the experiments considered here. Upon scattering on the crystal surface, the beam intensity as a function of time is monitored, and unfolding by the im-posed time structure reveals the change of the velocity distribution arising from the scattering process.

In the following, a short account will be given on the characteristics of the main components of a molecular beam scattering instrument. An example will be presented of a typical instrument to illustrate the problems arising at the level of a complete system.

6.3.1 Beam Sources

Undoubtedly, much of the recent progress in molecular beam spectroscopy can be attributed to the development of the free-jet beam source. A supersonic nozzle system provides a high intensity, nearly monochromatic beam and so combines the functions of a source and a monochromator. Some experimental setups designed for surface work do use monochromators either of the rotating disk type [7] or based on the crystal diffraction principle [8]. However, due to the particular attraction of mechanical simplicity, most of the attention to date is focussed onto the free-jet nozzle source [9].

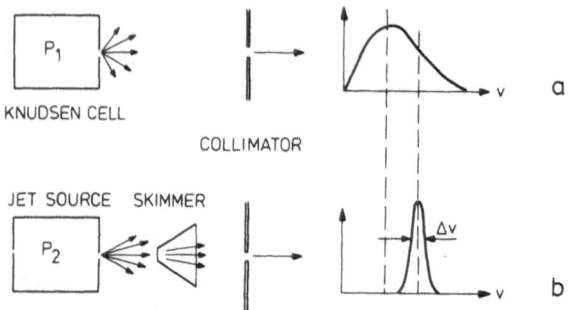

Fig.4 Comparison of a Knudsen cell (a) and a free jet nozzle source (b). The right hand diagrams indicate typical velocity distributions.

The principle of the nozzle beam source is best understood by comparison to an effusive source, the Knudsen cell. Here a gas at pressure P_1 is enclosed in a container with an orifice of diameter a (Fig.4a). Provided the orifice dimension

$$Kn = \lambda/a \geq 1 \tag{4}$$

is larger than unity, the flow through the orifice will be free molecular and all its properties can be calculated. In particular, the angular spread is a cosine distribution, and the velocity distribution is given by a Maxwellian stream as sketched in Fig.4a, with a most probable velocity

$$v_{mp} = (\frac{3}{2} \cdot \frac{2k_BT}{m})^{\frac{1}{2}} . \tag{5}$$

As the gas pressure in the container is increased, condition (4) does not hold any longer. The flow through the orifice becomes continuous, and a jet expansion behaviour prevails. As suggested by KANTROWITZ and GREY [10], most of the random gas motion is converted into forward directed motion, which results in a lobed angular distribution, and a narrow velocity distribution at the centerline of the lobe can be isolated by means of a conical skimmer (Fig.4b).

The most probable speed in a nozzle source is higher than that of an effusive source by a factor of $(2\gamma/3(\gamma-1))^{\frac{1}{2}}$. This amounts to about 30% for a monoatomic gas with $\gamma = 5/3$. As the velocity and thus the energy depend on the source temperature only, limits are given by the achievable range of practical temperatures T_s. An overview of characteristic beam properties for Helium and Neon is given in Table 2.

Table 2 Properties of nozzle beams

T_s [K]	Energy [meV]	v_{He} [cm/sec]	v_{Ne} [cm/sec]	k_{He} [Å^{-1}]	k_{Ne} [Å^{-1}]	λ_{He} [Å]	λ_{Ne} [Å]
20	4.31	$4.54 \cdot 10^4$	$2.03 \cdot 10^4$	2.9	6.5	2.18	0.97
70	15.1	$8.50 \cdot 10^4$	$3.80 \cdot 10^4$	5.4	12.1	1.16	0.52
300	64.6	$1.76 \cdot 10^5$	$7.87 \cdot 10^4$	11.2	25.0	0.56	0.25
3000	646.	$5.56 \cdot 10^5$	$2.49 \cdot 10^5$	35.3	78.9	0.18	0.08

The velocity distribution in a beam is conveniently described by the "speed ratio" $S = v/\Delta v$. This quantity, which determines the resolution in an inelastic experiment, depends on the Knudsen number [11]

$$S = \alpha \cdot Kn^{-(\gamma-1)/\gamma} \tag{6}$$

which is, for a monoatomic gas

$$S = \alpha \cdot Kn^{-0.4} \tag{7}$$

with a constant α depending on γ. Therefore, a high speed ratio can be achieved by increasing the gas pressure or the orifice diameter. Both these measures increase the total gas flow into the system, so the pumping capacity will be the ultimate limitation. The gas flow, however, increases with the square of the orifice dimensions, but only linearly with pressure. Therefore it is advantageous to work at a high pressure, combined with relatively small orifices.

The speed ratios presently achievable (with pressures of several hundred atmospheres and orifices of a few microns) are in a range of 200 [12,13] corresponding to a velocity resolution of 0.5% or an energy resolution of 1%. With a beam energy of 20 meV (LN$_2$ cooled source), this corresponds to a resolution of 200 μeV or about 1.6 cm^{-1}, which compares well to optical spectroscopic resolutions. This is not, as in most other spectroscopies, achieved by trading intensity. The nozzle beam source has the unusual and rewarding property of intensity and resolution increasing in parallel. The limit is, in any case, given by the pumping capacity of the vacuum system.

For systems where moderate resolution ($\sim$ 10%) can be tolerated, the alternative of a pulsed beam source can overcome the extreme pumping requirements. Here a fast switching valve combines the beam source, a limited monochromator and the time mark. Valves with pulse lengths of less than 20 μsec [14,15] have been developed. These are sufficiently compact to allow rotatable mounting inside a vacuum system. Intensities in the beam are about two orders of magnitude higher than with a continuous source, and the pulse repetition rate can be adapted to the pumping speed in the vacuum system.

6.3.2 Detectors

Neutral beams are by no means easy to detect. In most cases, the beam has to be ionized and then observed by either digital counting or analogue current measuring techniques. For dissociated molecules, a surface conductivity detector [16] is quite useful, however its response is not sufficiently fast for application in time-of-flight measurements. A similar limitation exists for detectors based on the low-temperature bolometer principle [17].

Provided signal intensities are large compared to the background gas density (which is not the normal case and occurs mainly for pulsed valve measurements), an ion gauge can be a suitable detector. Used in the Bayard-Alpert configuration with a central current collecting wire, its sensitivity is high and the response time is sufficient, provided the dimension of the ionization region is small compared to the flight distance. The limitation of this detector is usually given by the input noise of the current amplifier, whose bandwidth has to be compatible with the signal rise time. Extraction of the ions for preamplification in an electron multiplier partly overcomes this problem. Here extreme care is required in the ionizer design to avoid time delays in the ion extraction due to space charge effects.

As the scattered beam intensity becomes lower and comparable to the background gas level, mass spectrometric detection is usually required. Especially for beam atoms that are not components of the rest gas composition, like e.g. Helium, appreciable signal-to-noise ratio improvements can be achieved in this way. Commercial quadrupole mass filters have been used with success in this application, provided the ionizer is carefully designed for delay-free ion extraction. Both head-on and cross-beam ionizers can be suitable. Magnetic mass spectrometers have a better dispersion for low masses and are therefore preferable for detection of light gases like He or H_2.

In all cases differential pumping at the detector is a prerequisite for a good signal-to-noise ratio. This is not only necessary to reduce the background gas, but in particular for reduction of the residual level of the gas in the molecular beam. The latter component cannot be filtered out by a mass spectrometer and will therefore ultimately limit the achievable signal-to-noise ratios.

6.3.3 Systems

A schematic of the most successful system used to date for inelastic molecular beam investigations on surfaces [18] is shown in Fig.5. The beam source is a free jet nozzle operated with He at 200 At with a 5μ nozzle. The first stage is pumped with a massive 5000 l/sec diffusion pump. The chopper is a rotating disk at 150 Hz (9.000 Rpm) with a 0.2 mm slit to provide beam pulses of about 2 μsec length. The sample is mounted on a manipulator with 3 translational and 3 rotational degrees of freedom. The pressure in the sample chamber is in the 10^{-8} Torr region. An important detail is the 3-fold differential pumping in the flight tube between sample and detector. The latter consists of a head-on ionizer connected to a quadrupole mass filter or a magnetic deflection mass filter for minimum background at the Helium mass.

The velocity resolution of the system shown in Fig.5 is 0.8%. Operated with a LN_2 cooled nozzle at a beam energy of 18 meV this results in an energy resolution of 190 μeV. A typical spectrum (see for example Fig.14) is taken in 1-2 h.

From the short discussion presented here it is apparent that the most severe problem in a molecular beam system is the pumping requirements. The system described uses 9 pumping stages, which are required to maintain a Helium partial pressure differential of about 18 orders of magnitude between the nozzle (200 At) and the detector region ($\leq 10^{-12}$ Torr). This is necessary in order to obtain sufficient signal-to-noise ratio with a signal that is in-

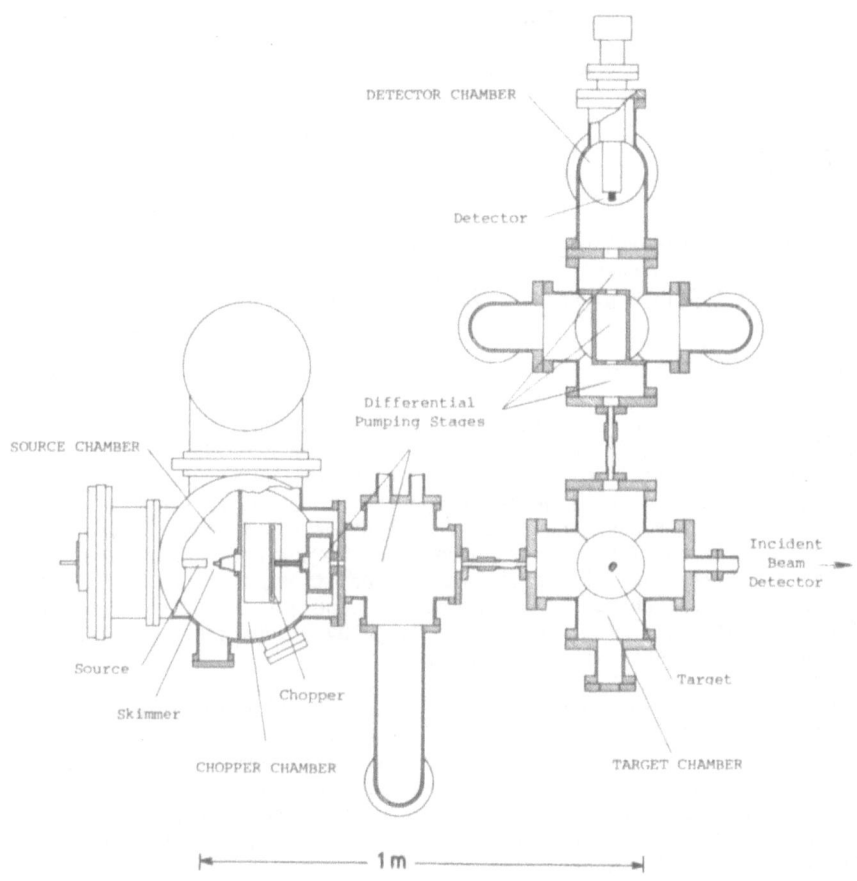

Fig.5 Apparatus for high-resolution inelastic molecular beam scattering from surfaces (courtesy of P.J. TOENNIES)

herently low. Typical elastic scattering intensities are in the percent scale, and inelastic peaks are found another 2 to 3 orders of magnitude lower in intensity. Another limitation arises from the mere physical dimensions of such a system, which make a design with freely variable incidence and scattering angle extremely difficult. Most instruments, therefore, rely on a fixed angle between source and detector, and a rotatable sample. This is not only due to the bulky pumps required, but also due to the flight distance, the minimum of which is essentially determined by the chopper speed and the data collection electronics.

6.4 Data Interpretation

6.4.1 Selection of Scattering Conditions

Inelastic scattering with the aim to measure the intrinsic vibration excitations of the surface is possible only if the experimental conditions are chosen such that quantum mechanical interaction is ensured. For example, bouncing a billiard ball from the surface will not tell us much about the surface phonon spectrum. As far as the incident atom or molecule is concerned, one would therefore choose a low mass and a low velocity, such that the wavelength of the particle is comparable to or larger than the surface periodicity. Also, it is desirable to have a gas-solid interaction potential with a shallow minimum. A gas that tends to chemisorptive interaction is therefore unsuitable. A good choice of the beam constituent is a low mass noble gas like Helium.

As a further consideration, we would like to interpret the measurement in terms of single-phonon events rather than many-phonon processes. A convenient quantity describing in a global way inelastic events is the Debye-Waller factor D,

$$I(T) = I_0 \cdot D \qquad\qquad D = \exp(-<(\underline{u} \cdot \Delta\underline{k})^2>) \qquad\qquad (8)$$

where $I(T)$ is an elastic (specular or diffracted) intensity, I_0 the intensity without inelastic attenuation, $\underline{u}$ is the thermal displacement of the surface atoms and Δk is the total momentum change of the gas atom, with $< >$ denoting the thermal average. A small Debye-Waller factor decreases the total inelastic intensity but increases the relative weight of single phonon events. The factor can be made small by either decreasing the surface temperature T (but the effect is limited due to the atomic zero-point oscillation), or by decreasing the momentum change. For the case of inelastic spectroscopy one is only free to modify the *normal* momentum change by either increasing the angle of incidence or decreasing the beam momentum.

A specific requirement on the surface under investigation is a long-range periodicity. Only this ensures the validity of the momentum conservation law parallel to the surface and therefore gives sense to terms like parallel momentum transfer and dispersion relations. It turns out from the above criteria that good single-phonon scattering conditions are expected for systems showing good diffraction properties in terms of intense, narrow diffraction maxima.

6.4.2 Selection Rules

Assuming now all conditions are chosen for single-phonon excitation in the quantum regime, it is important to notice a distinct difference between excitation of surface and bulk phonon modes. Bulk excitations will manifest themselves as continuum contributions to the spectra, since for each value of the parallel momentum component there may be a range of allowed states corresponding to different normal momenta. On the other hand, surface modes are uniquely defined by their parallel momentum component, so they contribute distinct narrow peaks to the spectra. This is the reason why a high resolution is a desirable property in IMBS.

Theories on the phonon excitation cross section in inelastic scattering [19-21] have been reviewed elsewhere [1]. Here we will only point out some qualitative features concerning excitation probabilities and selection rules. Atom-phonon coupling arises from the vibration induced change in the normal interaction potential, so, for a flat surface, only phonons with a polarization component normal to the surface are excited. This normal vibration selection rule becomes, however, less important for increasing surface corrugation. In contrast to spectroscopies relying on electromagnetic coupling (IR, LEELS), the forces in the atom-surface interaction act in phase on different atoms in the unit cell, coupling to acoustic excitations rather than optical ones. Again, this selection rule is not strict since the forces can be quite different in strength [22]. We also note that an acoustic branch will have an increasing ionic character as we proceed from the center of the Brillouin zone to its border. Consequently one would expect that the scattering cross section for acoustic phonons decreases as the border of the zone is approached.

6.4.3 Scattering Kinematics

The kinematic conditions for scattering are derived simply using energy conservation

$$E_f = E_i \pm \Delta E \tag{9}$$

and momentum conservation parallel to the surface

$$\underline{K}_f = \underline{K}_i + \Delta \underline{K} \quad . \tag{10}$$

For planar scattering, this leads to [23]

$$\frac{\Delta E}{E_i} = 1 - (1 - \frac{\Delta K}{K_i})^2 \frac{\sin^2 \Theta_i}{\sin^2 \Theta_f} \tag{11}$$

which forms a unique relation between energy and momentum transfer by a beam to a surface. It is usually displayed in the form of a coupling parabola as shown in Fig.6. Imagine a scattering experiment has been performed at a specific angle and incident energy leading to curve 1. To find the momentum transfer corresponding to a particular feature x in the time-of-flight spectrum, we calculate the energy transfer ΔE_x from the known flight distance and the incident beam energy. We can then use the coupling parabola to derive the corresponding momentum transfer ΔK_x.

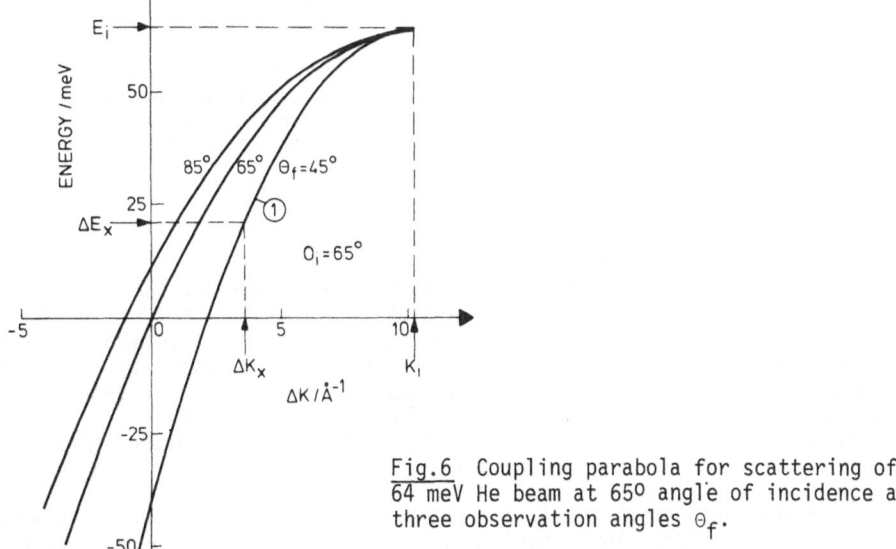

Fig.6 Coupling parabola for scattering of a 64 meV He beam at 65° angle of incidence and three observation angles Θ_f.

As one would like to compare the spectral features to phonon dispersion relations, it does not make sense to use momenta larger than the first Brillouin zone. Therefore the coupling parabola is folded back into the first zone using reciprocal lattice vectors as shown in Fig.7a. In the same diagram, one can now plot phonon dispersion relations, which are shown four times, namely for phonon creation (upper half) and annihilation (lower half), as well as for momentum gain and loss (right and left half, respectively). Any crossing point between the coupling parabola and the phonon dispersion indicates a condition where the beam with the parameters given can couple to the phonon spectrum, so a peak is expected in the time-of-flight spectra at the corresponding time. A further step is to use a reduced Brillouin zone where the phonon dispersion is plotted a single time and to fold the coupling parabola into the first quadrant, as shown in Fig.7b. The vertical coordinate now describes both energy gains and losses, which correspond to different parts of the coupling parabola as indicated by the full and dashed portions.

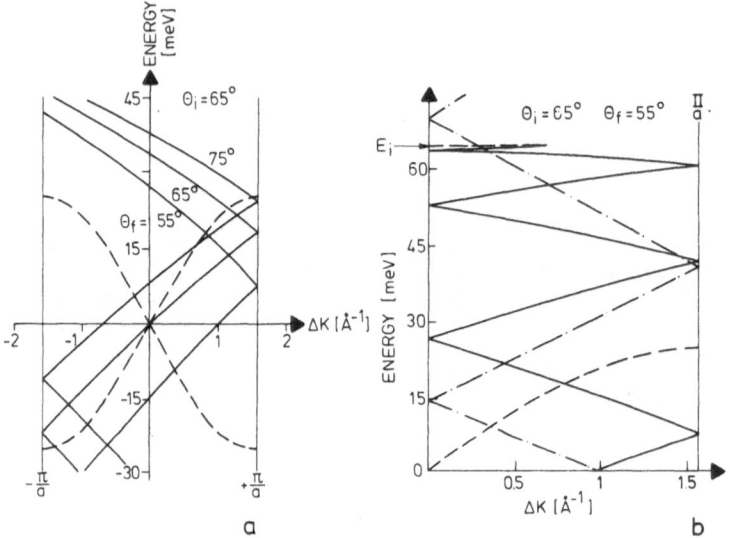

Fig.7 Coupling parabola shown in a single (a) or irreducible (b) Brillouin zone for incident conditions as in Fig.6. The dashed line represents the Rayleigh phonon dispersion. In the irreducible zone (b) the full line refers to phonon creation and the dash-dotted line to phonon annihilation.

6.5 Experimental Results

Inelastic effects have always been noticeable in gas-surface scattering, but frequently in a form not suitable for detailed interpretation. In particular, earlier experiments were performed in a region far from the quantum scattering regime considered here. In the so-called thermal regime, which is characterized by lobular rather than sharply directed scattering, substantial energy exchange between molecule and surface prevails. This is, however, usually connected with multiphonon effects and such does not lend itself to interpretation in terms of surface dynamical properties.

In measurements on diffractive systems, inelastic effects were mostly re-garded as a nuisance. Their influence is globally described by a Debye-Waller factor which includes multi-phonon, bulk, and surface phonon contributions. As the usefulness of single phonon measurements was pointed out [24] and calculated [19-22], attention first turned to the inelastic tails observable in the vicinity of diffraction peaks, where some structure could be identified without energy resolution. Techniques for the measurement of energy transfer were successively developed, up to a point where both energy and momentum analysis allowed the experimental investigation of surface phonon dispersion relations.

6.5.1 Inelastic Studies Without Energy Resolution

Early experiments set out to study the inelastic components in the vicinity of specular or diffraction peaks were performed on silver [25] and on alkali halides [26,27]. The latter measurements used out-of-plane scanning as illustrated in Fig.8a. A beam was impinging on a cleaved (100) crystal surface with a detector movable in the incident plane. Then the crystal was tilted by an angle $\delta\psi$ around the <10> direction such that the specular (or the in-plane diffracted) beam was not any more in the detection plane. Moving the detector then scanned the out-of-plane tail of the elastic peak, as shown in Fig. 8b for two tilt angles of a LiF surface held at 150 K. Side lobes are apparent in the data, with peak positions moving away from the elastic peak with increasing tilt angle. A calculation including a bulk Debye model was able to reproduce this behaviour [28], while an attempt to derive a dispersion relation for surface phonons from the data [27] was only partially successful.

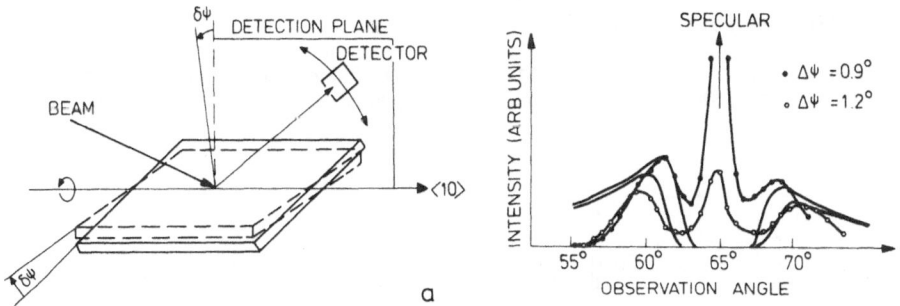

Fig.8 Observations of the inelastic tail near the specular peak by out-of-plane scanning (a). Measurements (b) are presented for two tilt angles $\Delta\psi$ on a LiF (100) surface at 150 K, $\Theta_i = 65^\circ$, <10> azimuth with a Helium beam of 60 meV (Ref. 26). Full lines are calculated results from Ref. 28.

An in-plane scan of the inelastic tail around a Helium beam scattered specularly from an epitaxially grown Ag(111) surface is shown in Fig.9a [25]. Here a nozzle beam with a variable binary gas mixture of He with Ar was used to produce He beams of variable speed or effective translation temperature as indicated for two curves in Fig.9a. Again, maxima belonging to "inelastic lobes" can be identified. A calculation based on a bulk Debye model [32] assuming single phonon interaction can reproduce the shape of the spectra, while the intensity of the inelastic contribution relative to the specular peak is not well represented.

A simple consideration can give an indication regarding the origin of the observed inelastic tails. Looking at Fig.6 and assuming phonon momenta small on the scale given, we notice that at observation angles larger than specular, say 85°, the curve cuts the vertical axis (corresponding to vanishing phonon momentum) in the upper half plane, while the opposite is true for observation angles between the specular and the surface normal direction. One expects therefore, as a rule of thumb, to find energy loss or phonon creation events for angles betwene specular and grazing, and energy gain or phonon annihilation contributing to scattering in directions between specular and normal. In Fig. 8b and Fig. 9a, inelastic lobes to the left of the specular position are therefore attributed to phonon annihilation, while phonon creation prevails at the right. As expected, annihilation processes are dominant for a metal surface observed well above the Debye temperature (Fig.9a), while comparable contributions from creation are apparent for LiF measured below the Debye temperature (Fig.8b).

More recent elastic scattering measurements obtained with high signal to noise ratio reveal a multiplicity of relatively sharp structures in the inelastic area between diffraction maxima [17,18]. This was explained theoretically in terms of inelastic scattering using the distorted-wave Born approximation, and some of the sharp features were related to van-Hove-type singularities arising from Rayleigh waves or Lucas modes [28]. Bound state resonances have also been demonstrated to have an important effect on the detailed angular structure of the inelastic intensity. For two systems, viz. He on LiF [29] and He on Graphite [30], such data have been obtained and interpreted in terms of surface phonon assisted bound state resonances.

6.5.2 Velocity Resolved Studies

Velocity measurements of inelastically scattered He atoms have been made using the phase shift information in a chopped beam, as shown in Fig.9b

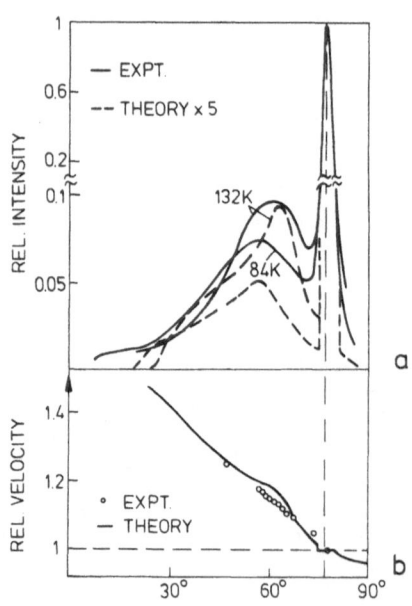

Fig.9 Inelastic tail near the specular peak for He scattering from an Ag(111) surface at 550 K (Ref. 25). Angular distribution for two beam temperatures (a) and velocity distribution for E_i/k_B = 132 K (b). Calculations from Ref. 32.

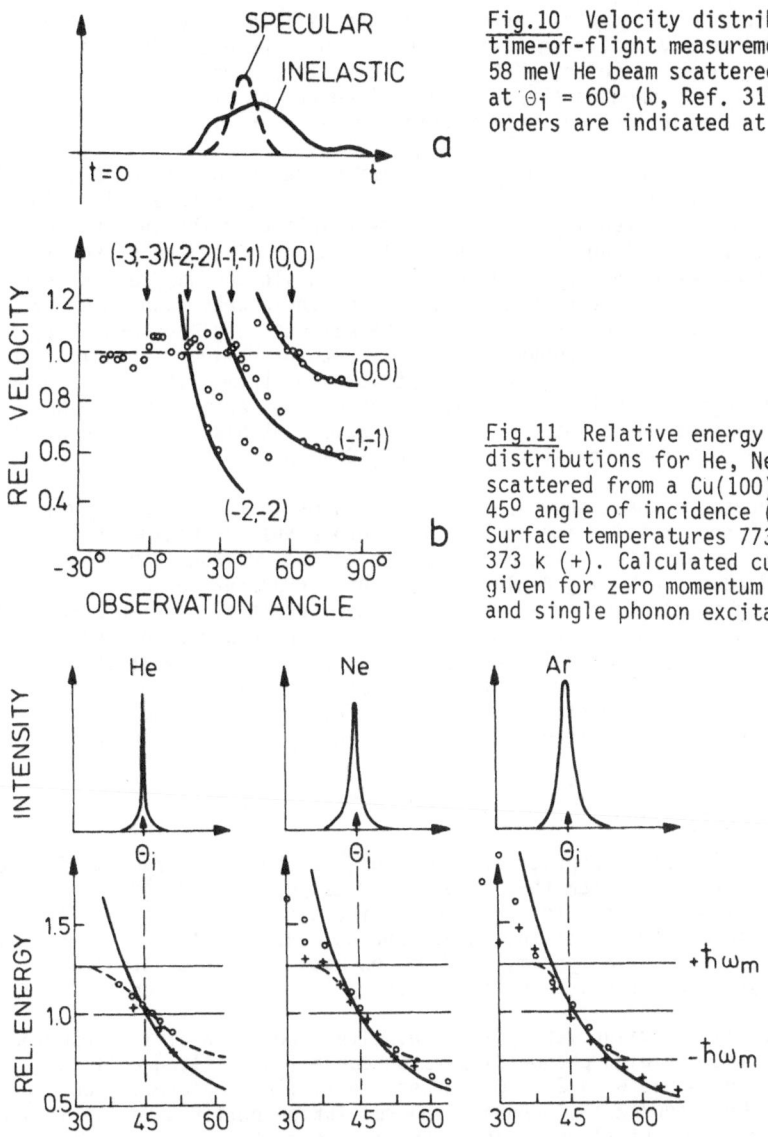

Fig.10 Velocity distributions from
time-of-flight measurements (a) for a
58 meV He beam scattered from LiF(100)
at Θ_i = 60° (b, Ref. 31). Diffraction
orders are indicated at the top.

Fig.11 Relative energy and intensity
distributions for He, Ne, and Ar
scattered from a Cu(100) surface at
45° angle of incidence (Ref. 33).
Surface temperatures 773 K (•) and
373 k (+). Calculated curves are
given for zero momentum transfer (−)
and single phonon excitation (−−).

for the Ag(111) surface. As expected from the rule of thumb given above,
velocity increases are found when observing toward the surface normal. The
amount of energy gain to the beam is appreciable, as indicated in Fig.9b,
where up to 20% velocity increase has been measured. In a diffractive system,
the inelastic lobes of the specular and the various diffraction peaks can
overlap. Under these circumstances, one may expect multiple structure in a
time-of-flight measurement, as indicated in the top part of Fig.10 [31].
The results for various observation angles, using a beam of 58 meV He atoms
impinging at 60° on a LiF (100) surface, are presented in the lower part of
the same figure. Each point indicates a maximum (or shoulder) in the time-

of-flight curves. The vertical arrows give the diffraction directions as indicated, and zero velocity shift is found at these points. The full lines represent the expected velocity shift according to a model assuming vanishing momentum transfer to phonons [32]. A general agreement is found, in particular for the area near low order diffraction peaks, where intensities are high.

Similar results have been found for inelastic scattering of He, Ne, and Ar from a (100) surface of copper [33] as shown in Fig.11. Here angular plots are presented of the intensity and the energy relative to the incident energy as a function of scattering angle. The full lines indicate the energy transfer predicted for zero phonon-momentum transfer, while the dashed line has been computed assuming single-phonon interaction and a Debye continuum [32]. The horizontal lines marked $\pm\hbar\omega_m$ indicate the maximum single-phonon energy transfer for phonons at the Debye energy. Except for Helium scattering, the data provide evidence for considerable contributions from multi-phonon processes. This suggests that the Cu(100)-Ne and Cu(100)-Ar systems are not within the limits of the quantum scattering regime, as also indicated by the lobular scattering characteristics, while Cu(100)-He probably is.

All data presented here show inelastic processes in the systems investigated, with phonon creation and annihilation processes dominating in separate angular ranges as expected. The measurements can be reproduced by phenomenological calculations assuming negligible momentum transfer to phonons or by a single-phonon model assuming a Debye continuum, eventually using a surface Debye temperature different from the bulk value. The measurements do not allow any discrimination of bulk versus surface effects, while it seems to be possible to extract some qualitative indication of single- versus multiphonon contributions. One experiment using a crystal energy analyzer [8] reports no evidence for contributions from surface phonons in the inelastic spectra.

6.5.3 Dispersion Relations

One of the most important aspects arising from the above measurements is the necessity to work in a system that is as close as possible to the limit of quantum interaction, if an interpretation of the data in terms of single phonon dispersion relations is intended. Recent measurements have therefore concentrated on the He-LiF system, where a low-mass inert gas atom and a highly corrugated, diffractive surface ensure proximity to the desired limit.

At time-of-flight measurement with a velocity resolution of 15% reported two points on the Rayleigh phonon dispersion curve [34] fitting well to a limiting speed of sound of $4.2 \cdot 10^5$ cm/sec. A similar measurement obtained with a pulsed beam source [35] and a speed resolution better than 10% is shown in Fig.12. Here time-of-flight tracks are given for the specular reflection ($\Theta_i = 67.5°$) and two off-specular positions obtained by rotating the sample crystal by $\pm 8°$. The peak positions of the inelastic curves are clearly shifted to opposite directions with respect to the elastic curve, with one spectrum showing an additional shoulder. These shifts have been observed for a range of angles, and Fig. 13 shows the accumulated results. As seen previously, energy gains and losses are separated in the angular range, except for the double structure shown in Fig.12 for +8°. The results are compared to the flight time shifts expected for single Rayleigh phonon interaction (full lines). Two branches arise from the fact that the coupling parabola cuts the Rayleigh phonon dispersion curve twice (see, e.g. Fig.7a, $\Theta_f = 65°$). Nearly all observed points are found on branch 1, which represents the case of smaller energy and momentum transfer. The reason for this behaviour

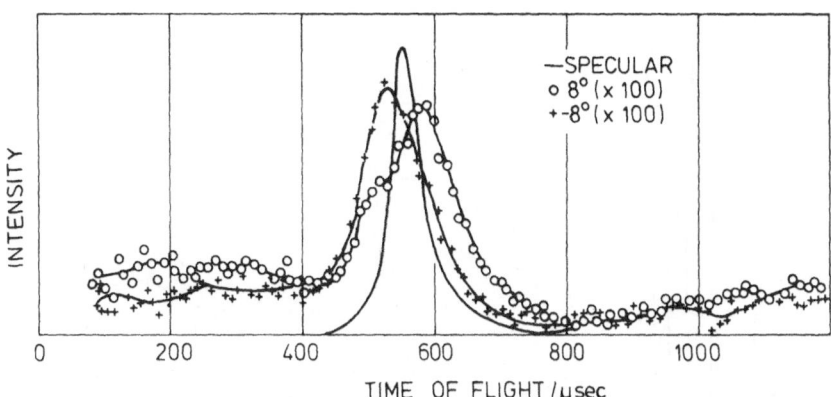

Fig.12 Time-of-flight curves for scattering of a 64 meV He beam from a pulsed source by a LiF(100) surface around 67.5° angle of incidence, for two sample positions ± 8° from the specular condition (Ref. 35).

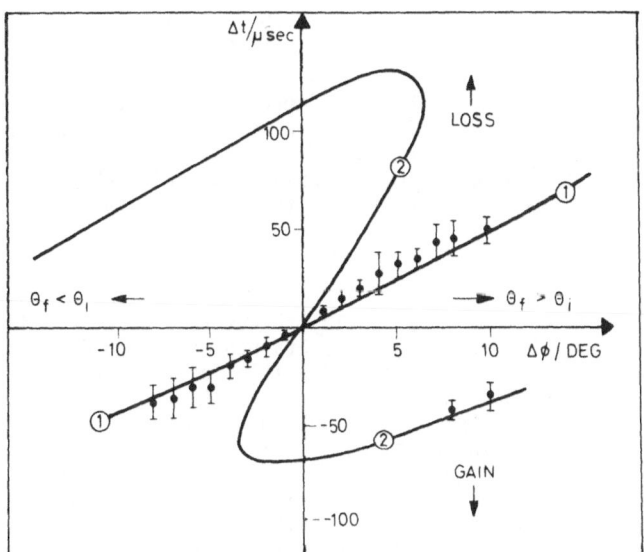

Fig.13 Time-of-flight peak positions (see Fig.12) for various observation angles (Ref. 35). Full lines: calculated positions for single Rayleigh phonon interaction.

arises from a rapid decrease of intensity with energy transfer, which will always favour branch 1 in a measurement that does not resolve the two branches. Only if the intensity in branch 1 has dropped to a level comparable to the other branch, a double peak structure is observable as shown for the +8° case. It should be noted that the two points reported in an earlier experiment [34] fall on branch 2, for reasons not apparent to the author.

The measurements presented are indicative of an important contribution of single Rayleigh phonons to the inelastic scattering in the He-LiF system, however, they are by no means conclusive. The high-resolution results recently

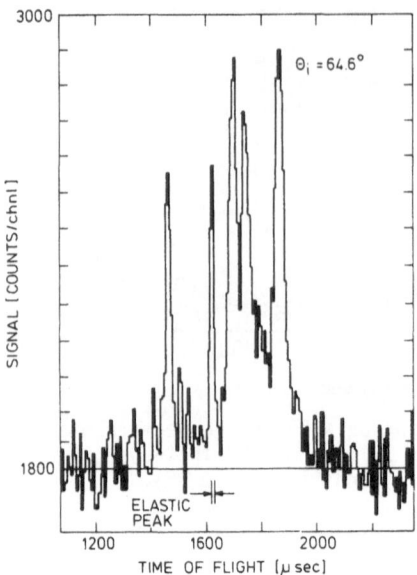

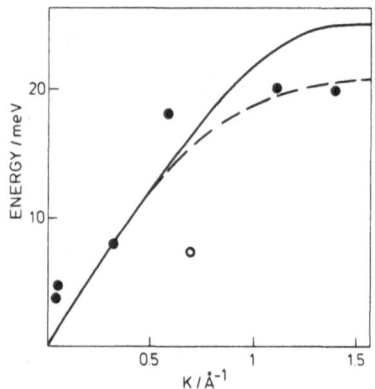

Fig.14 High-resolution time-of-flight measurements for a 18 meV He beam scattered from a LiF(100) surface (Ref. 18). Angle between beam source and detector. is 90°, (Courtesy of J.P. TOENNIES).

Fig.15 Energy and momentum transfer from time-of-flight measurements for He on LiF(100) (Ref. 36,18), compared to a calculated Rayleigh phonon dispersion with (--) and without (-) relaxation (Ref. 37).

obtained by BRUSDEYLINS, DOAK and TOENNIES [36,18] therefore present a major breakthrough. A typical time-of-flight spectrum obtained with a 18 meV He beam on a LiF(100) surface is shown in Fig.14. A multiplicity of extremely sharp and well-defined peaks is apparent. An early evaluation of these data is presented in Fig.15. Here the dots indicate the position in energy-momentum space of the sharp features as shown in Fig.14. They are compared to a dispersion relation of surface Rayleigh phonons [22] (full line) of an un-relaxed surface, and to a modified dispersion relation taking into account a surface polarizability different from the bulk value (dashed line, Ref. 37). The measured points correlate well with the lower calculated curve, except for the open circle. Recent measurements (B. DOAK, private communication) indicate that the scatter in the full dots arises from a slight misalignment of the azimuth angle and can be drastically reduced. In contrast, the open point seems to be part of a yet unidentified separate branch.

The recent high-resolution data lead to a number of important deductions. First, they reveal that earlier results just represent an envelope over multiple peaks, as suspected. Second, the sharp structure found is indicative of surface mode interaction rather than bulk interaction (see 4.2). In fact, no significant bulk contribution to the spectra could be identified. Third, it is apparent that Rayleigh phonons right in the vicinity of the Brillouin zone boundary can be detected in contrast to an expected selection rule. The influence of bound state resonances on the inelastic intensities seems to dominate the decay of coupling strength near the zone boundary.

6.6 Conclusions and Outlook

Inelastic Molecular Beam Spectroscopy on surfaces has now reached a point in its development as an experimental method where it changes from a "promising technique" to a laboratory tool. A number of important foundation stones have been laid, on which experiments leading potentially to significant progress in surface physics can be built.

It has been demonstrated that IMBS is capable of measuring excitation *frequencies* of surface localized vibration modes with a resolution comparable to optical methods. Such mode frequencies are observable through energy ranges inaccessible to electron spectroscopy, namely for small transition energies where LEELS spectra are swamped by the strong elastic peak. The sensitivity of IMBS to acoustic vibrations makes it complementary to both electron and photon spectroscopies. In addition, it is complementary to these methods in that it can access vibration modes well removed from the center of the Brillouin zone.

IMBS can also measure the *momenta* or wave numbers of surface localized vibration modes, and can access these for a range covering the whole Brillouin zone. This allows to determine uniquely dispersion relations for surface modes, in complete analogy to the mapping of bulk dispersion relations obtained by neutron spectroscopy. The data presently available indicate that such observations are possible right up to the border of the Brillouin zone, where short-range interactions dominate the dispersion behaviour and where therefore surface specific differences to the bulk properties should have the strongest influence.

So far, the above properties have only been demonstrated for one single case, namely the observation of Rayleigh-type surface phonons on an ionic crystal surface. This leaves, of course, a number of questions on the applicability of IMBS open. First, one would like to see whether coupling to *optical phonons* is possible. This could give access to an investigation of the interaction forces within the unit cell. While, as mentioned in 4.2, a selection rule seems to forbid observation of optical modes, the mere fact that acoustic modes are seen strongly at the zone boundary indicates they should be accessible, too. Second, the question of single-phonon interaction on *metal surfaces* is still open. Metals tend to have low Debye temperatures and show high Debye-Waller factors, and the corrugation amplitudes are considerably smaller than those found in ionic crystals. These facts seem to make it difficult to allow a proper choice of experimental parameters ensuring quantum interaction. Elastic measurements, however, have shown that metal surfaces can well act as diffractive systems [5,38], so inelastic single-phonon measurements should also be possible. Third, the exciting area of *adsorbate vibrations* and dispersion relations is completely open. Here again, exciting results in the elastic scattering of molecular beams are available [5,6], so we are looking forward to similarly exciting data in the realm of inelastic spectroscopy. Such measurements hold promise of giving access to important problems as adatom-adatom interaction, surface reconstruction and the whole range of soft-mode phenomena including two-dimensional phase transitions.

Acknowledgements

The author is indebted to Dr. E.A. Trendelenburg, Dr. D.E. Page, and Dr. B. Fitton for generous support and continued interest in this work. He gratefully acknowledges interesting discussions with Prof. J.P. Toennies, Prof.

G. Boato, Dr. G. Benedek and Dr. B. Doak, who have guided him with patience into this field. Particular thanks go to L. de Boer for her help in the preparation of the manuscript.

References

1) F.O. Goodman and H.Y. Wachman, "Dynamics of Gas-Surface Scattering", Academic Press, New York, 1976.
2) J.P. Toennies, Appl. Phys. 3, 91 (1974).
3) For a recent review see: H. Wilsch in "Topics in Surface Chemistry", ed. by E. Kay and P.S. Bagus, Plenum, 1978, p. 135.
4) M.J. Cardillo and G.E. Becker, Phys. Rev. Letters 40, 1148 (1978).
5) K.H. Rieder and T. Engel, Phys. Rev. Letters 43, 373 (1979).
6) K.H. Rieder and T. Engel, Phys. Rev. Letters (to be published).
7) P. Scherb and E.M. Hörl, Rev. Sci. Instr. 47, 1511 (1976).
8) B.F. Mason and B.R. Williams, Surface Sci. 77, 385 (1978).
9) J.B. Anderson in "Molecular Beams and Low Density Gas Dynamics", ed. by P.P. Wegener, M. Dekker, New York 1974, p. 1.
10) A. Kantrowitz and J. Grey, Rev. Sci. Instr. 22, 328 (1951).
11) J.B. Anderson and J.B. Fenn, Phys. Fluids 8, 780 (1965).
12) B. Brusdeylins, H.D. Meyer, J.P. Toennies, and K. Winkelmann, Progr. Astronaut. Aeronaut. 51, 1047 (1977).
13) R. Campargue, A. Lebéhot, and J.C. Lemonnier, Progr. Astronaut. Aeronaut. 51, 1033 (1977).
14) W.R. Gentry and C.F. Giese, Rev. Sci. Instr. 49, 595 (1978).
15) M.R. Adriaens, W. Allison, and B. Feuerbacher, to be published.
16) H. Nahr, H. Hoinkes, and H. Wilsch, J. Chem. Phys. 54, 3022 (1971).
17) G. Boato, P. Cantini, and L. Mattera, Surface Sci. 55, 141 (1976).
18) B. Doak and P. Toennies, private communication.
19) N. Cabrera, V. Celli, F.O. Goodman, and R. Manson, Surface Sci. 19, 67 (1970).
20) G. Wolken, J. Chem. Phys. 58, 2047 (1973).
21) M. Lagos, Surface Sci. 65, 124 (1977).
22) G. Benedek and G. Seriani, Japan. J. Appl. Phys. Suppl. 2, Pt. 2, 545 (1974).
23) G. Benedek, Phys. Rev. Letters 35, 234 (1975).
24) N. Cabrera, V. Celli, and R. Manson, Phys. Rev. Letters 22, 346 (1969).
25) R.B. Subbarao and D.R. Miller, J. Chem. Phys. 51, 4679 (1969); J. Vac. Sci. Technol. 9, 808 (1972).
26) B.R. Williams, J. Chem. Phys. 55, 3220 (1971).
27) B.F. Mason and B.R. Williams, J. Chem. Phys. 61, 2765 (1974).
28) F.O. Goodman and W.K. Tan, J. Chem. Phys. 59, 1805 (1973).
29) P. Cantini, G.P. Felcher, R. Tatarek, Phys. Rev. Letters 37, 606 (1976).
30) P. Cantini and R. Tatarek, private communication (1980).
31) S.S. Fisher and J.R. Bledsoe, J. Vac. Sci. Technol. 9, 814 (1972).
32) F.O. Goodman, J. Vac. Sci. Technol. 9, 812 (1972); Surface Sci. 30, 1 (1972).
33) J. Lapoujoulade and Y. Lejay, J. Chem. Phys. 63, 1389 (1975).
34) J.M. Horne and D.R. Miller, Phys. Rev. Letters 41, 511 (1978).
35) B. Feuerbacher, M.R. Adriaens, and H. Thuis, Surface Sci. 94, L171 (1980).
36) G. Brusdeylins, R.B. Doak, and J.P. Toennies, Phys. Rev. Letters (to be published).
37) G. Benedek and N. Garcia, Phys. Rev. Letters (to be published).
38) G. Boato, P. Cantini, and R. Tatarek, J. Phys. F 6, L237 (1976).

7. Neutron Scattering Studies

C. J. Wright

With 6 Figures

Neutrons, neutral particles scattering weakly from the nuclei of atoms, are most commonly encountered as probes for the bulk behaviour of materials. However, they have a number of properties which can be exploited to advantage in surface characterisation.

Neutrons are scattered relatively strongly by hydrogen and so the doping or the covering of a high-surface-area material with hydrogen atoms transforms the technique into a surface-sensitive one. Combining this scattering enhancement and surface sensitivity with the capability of the neutron to penetrate materials, leads to the possibility of observing surface vibrations under conditions of high temperature and pressure, something which no other technique is capable of.

Since the scattering is from nuclei it is describable by simple theory in which the intensity of scattering from a particular adsorbent-absorbate complex can be described solely in terms of the atomic masses and the interatomic force constants. This again distinguishes the technique from others and the exploration of these two features will probably account for most of the use of neutrons in studying surface vibrations in the future.

A third factor of importance is the mass and consequently the relatively large momentum of a thermal neutron. This makes possible the exploration of excitations with momentum transfers of the order of reciprocal lattice vectors. Collective surface excitations and inter-adsorbate force constants may thus be determined and as an example the scattering from argon adsorbed on carbon [1] will be described later.

A summary of the factors which differentiate the neutron technique from others described in this book is given below

	Comments
No selection rules	All vibrations involving displacements of strongly scattering atoms are visible
Low absorption cross sections	Enables "in-situ" determinations at high temperatures and pressure
Low neutron fluxes	Leads to the requirement that samples must have high specific surface areas like those of oriented graphites or metal powders. N.B. This can be an advantage since electron-energy-loss spectroscopy is not presently applicable to powders.

The neutron's mass	Excitations with momentum transfers comparable to reciprocal lattice vectors can be determined, so leading to interadsorbate force constants.
Large Debye-Waller factors	The present generation of spectrometers only allow high energy transfers to be observed with high associated momentum transfers. This results in large Debye-Waller factors which make the observation of scattering at energies >3000 cm^{-1} relatively difficult.

7.1 Theoretical Background

The scattering from an isotropic harmonic oscillator can be derived [2,3] from the correlation function representation of the scattering law

$$S(Q,\omega) = \frac{1}{2}\,\pi\hbar \int_{-\infty}^{+\infty} dt\ \exp(-i\omega t) <\exp[-iQ\cdot r(0)]\exp[iQ\cdot r(t)]> \tag{1}$$

where Q and ω are the respective momentum and energy transfers incurred on scattering, and where r is the displacement of the scattering nucleus from its equilibrium position.

Since $r(t)$ can be written

$$r(t) = [\hbar/(2M\omega_0)]^{\frac{1}{2}}[a\ \exp(-\omega_0 it) - a^+ \exp(i\omega_0 t)] \tag{2}$$

where a and a^+ are the annihilation and creation operators, then after substitution it can be shown that

$$S(Q,\omega) = \exp\left[\frac{-\hbar Q^2}{2M\omega_0}\coth\left(\frac{\hbar\omega_0}{2kT}\right)\right]\ \exp\left(\frac{\hbar\omega}{2kT}\right) \sum_{n=-\infty}^{\infty}\ \delta(\hbar\omega - n\hbar\omega_0)$$

$$\times\ I_n\left[\left(\frac{\hbar Q^2}{2M\omega_0}\right)\mathrm{cosech}\left(\frac{\hbar\omega_0}{2kT}\right)\right]\quad . \tag{3}$$

In (3) the various possible excitations are given by $\omega = \sum_n n\omega_0$ (n is positive for energy loss) and I_n is the modified Bessel function of the first kind.

The cross-section, or the intensity of scattering per incident neutron per steradian per unit energy transfer, is

$$\frac{d^2\sigma}{d\Omega dE} = \frac{Nk}{k_0}\,\frac{\sigma}{4\pi}\,S(Q,\omega) \tag{4}$$

where k and k_0 are the final and incident neutron wave vectors, and σ is the atomic cross-section. The atomic cross-sections most frequently encountered

---

$\hbar$ = h/2π (normalized Planck's constant).

Table 1. Neutron scattering and absorption cross-sections of some atoms commonly encountered in surface science

	$\sigma_{inc} + \sigma_{coh}$	σ_{coh}	σ_{abs}
H	81.5	1.76	0.19
D	7.6	5.59	0.0005
C	5.51	5.56	0.003
N	11.4	11.1	1.1
O	4.24	4.23	0.0001
Al	1.5	1.54	0.13
Si	2.2	2.22	0.06
S	1.2	0.99	0.28
^{36}Ar		74.1	
Fe	11.8	11.34	1.4
Co	6	0.79	21
Ni	18.0	13.33	2.7
Mo	6.1	5.98	1.4
Pd	4.8	4.52	4.0
Pt	12	11.34	5.0

in surface science are recorded in Table 1 [4] which includes the total scattering cross-section $\sigma_{inc} + \sigma_{coh}$, of importance in most of the work in this chapter, and σ_{coh}, the coherent cross-section, of importance in connection with the observation of scattering from collective excitations.

The simple theory of (3) has to be modified for a surface oscillator to take account of its anisotropy. Expressing the scattering law in terms of the intermediate scattering functions, it can be written as the integral of a product, if it is assumed that the parallel and perpendicular vibrations do not interact with each other

$$S(Q,\omega) = \frac{1}{2\pi\hbar} \int_{-\infty}^{+\infty} \exp(-i\omega t) <I_A(Q,t) \cdot I_E(Q,t)>_\theta \, dt \tag{5}$$

After a derivation similar to the one above, and averaging over all orientations of the crystallites with respect to the scattering vector, the following equations are obtained for the intensity [5]. For energy transfer $\hbar\omega_A$

$$\exp\left(\frac{\hbar\omega_A}{2kT}\right) \int_0^1 d\lambda \, \exp -\left[\frac{\hbar Q_A^2 \lambda^2}{2M\omega_A} \coth\left(\frac{\hbar\omega_A}{2kT}\right) + \frac{\hbar Q_A^2}{2M\omega_E}(1-\lambda^2) \coth\left(\frac{\hbar\omega_E}{2kT}\right)\right]$$

$$\times \, I_1\left[\left(\frac{\hbar Q_A^2}{2M\omega_A}\right)\text{cosech}\left(\frac{\hbar\omega_A}{2kT}\right)\lambda^2\right] I_0\left[\left(\frac{\hbar Q_A^2}{2M\omega_E}\right)\text{cosech}\left(\frac{\hbar\omega_E}{2kT}\right)(1-\lambda^2)\right] \, , \tag{6}$$

for energy transfer $\hbar\omega_E$

$$\exp\left(\frac{\hbar\omega_E}{2kT}\right) \int_0^1 d\lambda \, \exp -\left[\frac{\hbar Q_E^2}{2M\omega_A} \coth\left(\frac{\hbar\omega_A}{2kT}\right)\lambda^2 + \frac{\hbar Q_E^2}{2M\omega_E} \coth\left(\frac{\hbar\omega_E}{2kT}\right)(1-\lambda^2)\right]$$

$$\times \, I_1\left[\left(\frac{\hbar Q_E^2}{2M\omega_E}\right)\text{cosech}\left(\frac{\hbar\omega_E}{2kT}\right)(1-\lambda^2)\right] I_0\left[\left(\frac{\hbar Q_E^2}{2M\omega_A}\right)\text{cosech}\left(\frac{\hbar\omega_A}{2kT}\right)(1-\lambda^2)\right] \, , \tag{7}$$

with similar expressions for harmonics and combination bands of these fundamentals.

The scattering intensity at a particular energy is the sum of the intensities of all the individual excitations whose net energy transfer is equal to the energy in question. This theory is sufficient to describe the scattering from an oscillator bound to a substrate of infinite mass. In real systems the substrate's vibrations interact with those of the adsorbate so that the above equations need further modification before they can be used to describe satisfactorily experimental data. In particular, they predict scattering intensities for individual excitations which are too large, because they neglect the component of the amplitude of vibration due to the substrate's motion. This reduces the total intensity through a term in the scattering law analogous to a Debye-Waller factor.

The simplest way to include this factor into the formulation of the scattering law is to represent the substrate vibrations by an additional Einstein oscillator term in the intermediate scattering law

$$I(Q,t) = I_A(Q,t)I_E(Q,t)I_L(Q,t) \quad . \tag{8}$$

Following through the derivation of the scattering intensity in the same way, leads to an expression in which (6,7) are multiplied by an additional exponential term $\exp[(-\hbar Q^2/2M\omega_L) \coth(\hbar\omega_L/2kT)]$ for $\hbar\omega_L > 2kT$. This is a condition generally satisfied when the sample is cooled to 80 K or below. It has been shown that such an expression reproduces well the relative intensities of the excitations of atoms adsorbed at interfaces, if $\hbar\omega_L$ is treated as a variable.

It can be shown that at 0 K the mean square amplitude of vibration of an atom, in the rigid-ion approximation, is inversely proportional to the square root of its mass. In the limit of temperatures greater than the Debye temperature, however, the amplitudes for all atoms on a lattice whatever their mass tend to become equal Eq.(8), and the cross section derived from it, have been used to interpret spectra for $H_{0.5}TaS_2$ [6] where the hydrogen atoms are bound to the metal atoms, and for $H_{0.04}WS_2$ [7] where the bonding is via the sulphur atoms. Measured values for $\hbar\omega_L$ in these two cases are almost identical, suggesting that the high temperature approximation is applicable even at 80 K.

The interaction between an adsorbate and its adsorbent also leads to an enhancement of the scattering intensity from the adsorbent's surface vibrations. A displacement of the adsorbent atoms leads necessarily to a damped displacement of the adsorbate, with the magnitude of the damping determined by the force constant between the two atoms. Through this mechanism it was hoped that it would be possible to observe the vibrations of surface adsorbent atoms, and to identify whether these vibrations were different to those of atoms in the bulk. However comparisons of the density of states of bulk nickel [8] and platinum [9], with those of surface atoms revealed by the scattering from adsorbed hydrogen, did not show any differences.

On the other hand, substantial changes have been observed in the substrate density of states of molybdenum sulphide [10], in measurements made on material with different amounts of sorbed hydrogen. Figure 1 shows the apparent increase in energy of the peaks in the amplitude weighted density of states on hydrogen sorption. Two possible causes of this effect have been proposed. The first suggests that it arises because of force constant changes in the

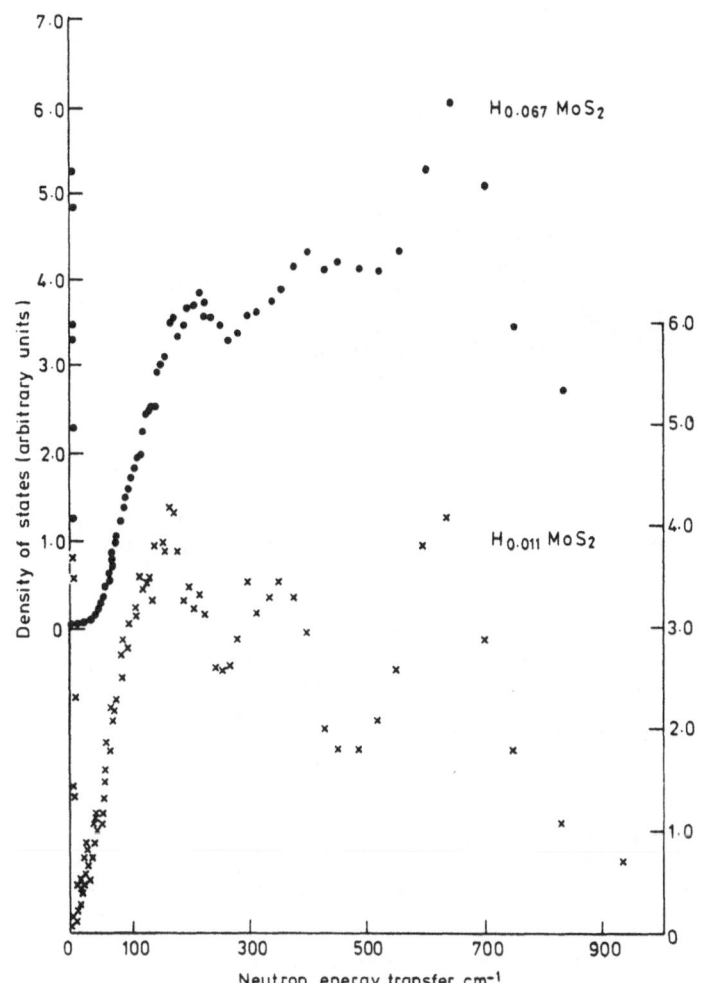

<u>Fig. 1.</u> Amplitude weighted density of states of $H_{0.011}MoS_2$ (×) and $H_{0.067}MoS_2$ (•)

substrate arising from electronic changes induced by the sorbate: the second that it is due to changes in the weighting of the different components of the density of states in the observed spectra. If the hydrogen should sorb specifically onto sulphur atoms, and if the sulphur atoms have higher amplitudes at higher frequencies than the molybdenum atoms, then the increased effective scattering cross-section of the sulphur hydrogen combination after sorption could cause an effect similar to that observed.

Experimental Considerations

Two types of neutron spectrometer are used to determine vibrational energy transfers. One, used essentially between 0 and 600 cm^{-1}, uses neutron time-of-flight techniques where a monochromated pulsed beam of low energy neutrons is up-scattered (neutron energy gain) by a sample and the energy transfers

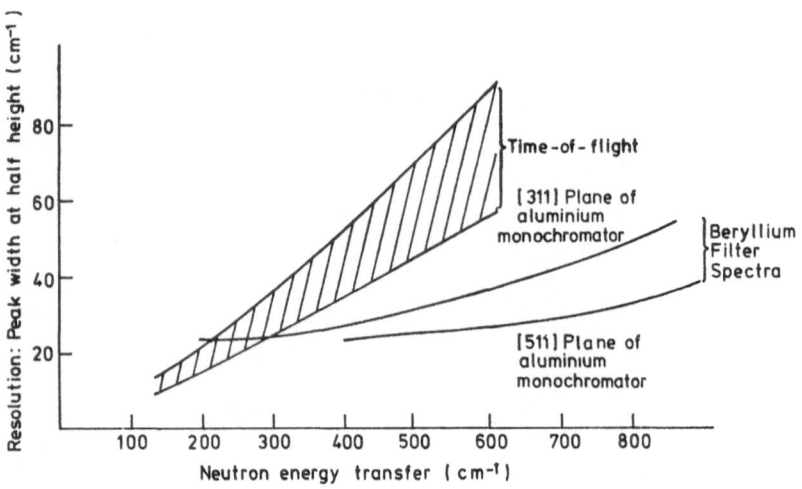

Fig. 2. Energy resolution of time-of-flight and beryllium filter spectrometers

are determined by the neutron times-of-flight over fixed sample-to-detector distances [11]. In these experiments the energy resolution $\Delta E/E_{transfer}$ is approximately constant over the energy-transfer range.

In the second type of experiment, known as beryllium-filter spectroscopy, the incident neutrons are monochromated by Bragg diffraction, down scattered (neutron energy loss) by a sample and then detected on the far side of a beryllium filter. This filter is constructed of blocks of polycrystalline beryllium interleaved with cadmium sheet such that all neutrons which will satisfy the Bragg relationship are diffracted by the beryllium and absorbed by the cadmium. Only those neutrons, for which $\lambda > 2d$ (i.e., > 4 Å), will be transmitted through the filter and detected. For this spectrometer the resolution is a sum of two components one (σ_w) due to the energy width of the window, and the other (σ_i) due to the energy resolution of the incident beam, with

$$\sigma_i = 0.851 \, E_i \, \cot(\theta)d\theta$$

where 2θ is the angle of diffraction from the monochromator and $d\theta$ is the error due to collimator width and monochromator mosaicity [12]. In general σ_w (~ 21 cm^{-1}) is the dominant contribution at low energies whereas σ_i dominates at higher energies. Usual conditions of operation result in resolutions $\Delta E/E$ of $\sim 10\%$.

A diagram showing typical magnitudes of the resolution, and the regions of energy transfer where the different spectrometer resolutions overlap is shown in Fig.2. It should be emphasised that, by optimising spectrometers for particular objectives, a wide variation in intensity and resolution can be obtained. Fig.2 indicates the relative performances of the different types of spectrometer, and can be used to estimate the resolution of many of the spectra included in this chapter.

There are a number of comprehensive reviews of the experimental aspects of neutron scattering, and the reader is recommended to them for further information [13,14].

7.2 Applications

All the techniques described in this book can be applied to the measurement of the vibration frequencies of adsorbed gases and to the identification of surface species. With neutron scattering it is particularly easy to do this both for hydrogenous materials and at low frequencies, when all vibrations in which hydrogen atoms are displaced will be observable.

Rather than reproduce here what will be covered elsewhere, the main emphasis of the applications discussed in this chapter will be on the ways in which the *intensity* information present in a neutron-scattering spectrum has enabled experimenters to derive structural information. It is this extra parameter that in the main has differentiated neutron scattering from all other techniques in the study of surface vibrations. Some important examples of the application of neutron scattering to frequency measurement are simply referred to below: C_2H_4 adsorbed by zeolite 13 X [16], hydrogen adsorbed by palladium

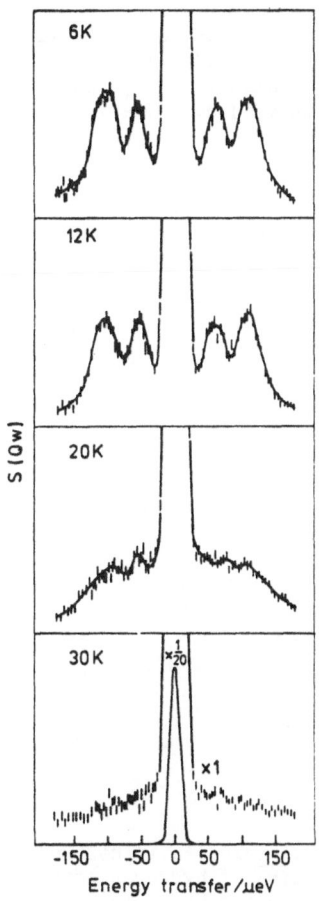

Fig. 3. Tunnelling spectra of 0.7 layers of CH_4 on Vulcan III at various temperatures

117

black [17], Raney nickel [18,19] and platinum black [20]. In addition, the wide energy range of neutron scattering measurements should be emphasised by pointing out that it has been possible to observe the tunnelling transitions of adsorbed methane on a graphite surface [21]. Figure 3 shows transitions at 5.8 µeV (0.468 cm^{-1}) and 108 µeV (0.872 cm^{-1}) and their progressive broadening on raising the temperature.

7.2.1 Sulphide Catalysts

Hydrodesulphurisation catalysts are materials commonly based upon mixtures of molybdenum sulphide (MoS_2), and cobalt sulphide (Co_9S_8), or tungsten sulphide (WS_2) and nickel sulphide (Ni_3S_2). These materials take up considerable quantities of hydrogen especially under the high-pressure ($50 \rightarrow 200$ at), high-temperature ($350 \rightarrow 420°C$) conditions in which they are commonly used. The vibration frequencies of this hydrogen are observable with inelastic neutron scattering but analysis of the intensity of the scattering provides additional information on the likely bonding of this hydrogen to the lattice. Figure 4 shows the measured spectrum from hydrogen adsorbed on tungsten sulphide at one atmosphere with, for comparison purposes, the spectrum of electrochemically prepared $H_{0.5}TaS_2$. Both spectra show a number of excitations at regularly spaced energy intervals, which are due to various fundamentals and their harmonics.

The frequencies of the excitations in the spectra are basically similar and, unless the intensity information had also been available, incorrect structural conclusions could have been deduced. The intensities, normalised to the intensity of the lowest energy peak, were analysed by comparing them

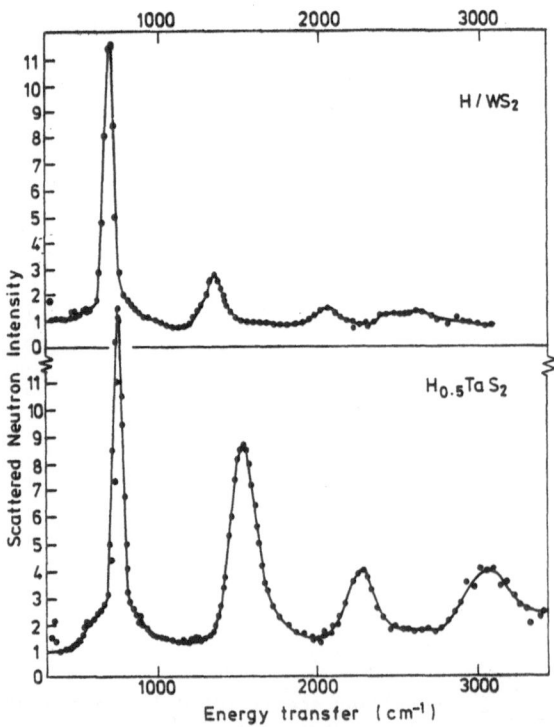

Fig. 4. Inelastic neutron spectra of hydrogen adsorbed by WS_2 and $H_{0.5}TaS_2$

with the predictions of (8) for a number of structural models. Three models were considered, (a), (b), and (c). In model (a), the hydrogen atoms lay in the planes of their neighbours (tantalum or sulphur atoms), and were multiply bonded to them. In models (b) and (c), the hydrogen atoms lay above the planes of their neighbours and were multiply, (b), or singly, (c), bonded to them. In addition in models (a) and (b) it was necessary to assume that there was approximate degeneracy between the highest-energy fundamental and the first harmonic of the lowest-energy fundamental. Only in this way could the number of observed bands be equated to the number predicted.

Since $\hbar\omega_L$ is unknown for the lattices in question, then if predictions are made on the assumption that it is zero, (8) shows that $\ln(I_{obs}/I_{pred})$ will be proportional to Q^2, where I_{obs} and I_{pred} are the observed and predicted neutron-scattering intensities, respectively. In Fig.5 this ratio is plotted

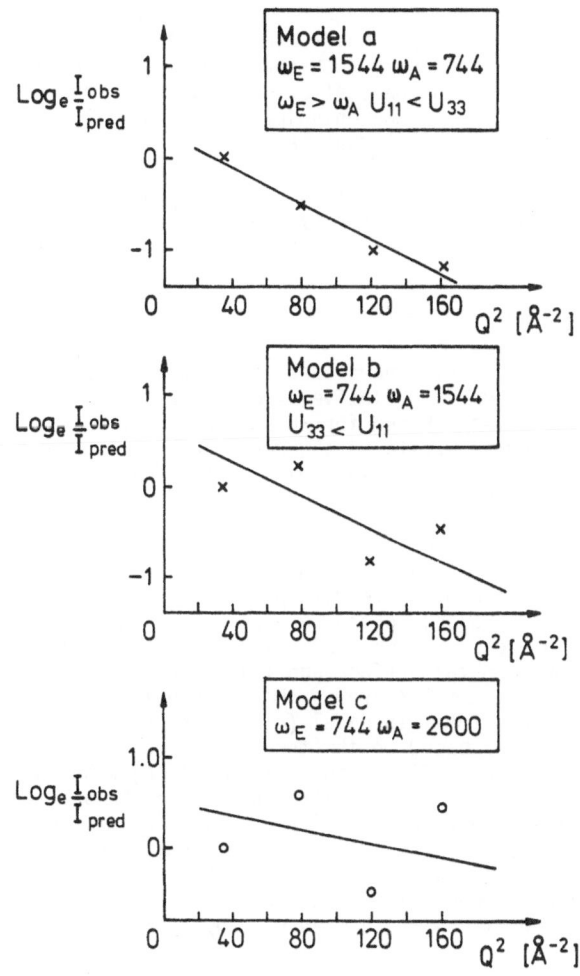

Fig. 5. $\ln(I_{obs}/I_{pred})$ plotted against Q^2 for the structural models of $H_{0.5}TaS_2$ (a), (b), and (c)

against Q^2 for $H_{0.5}TaS_2$, and it is apparent that only for model (a) do the points fall on a satisfactory straight line. This is in agreement with the interpretation of diffraction data taken from this material [6], which shows the hydrogen atoms to occupy sites in the centre of a triangle of metal atoms in the tantalum plane. In contrast, similar comparisons for hydrogen adsorbed on tungsten sulphide lead to a different conclusion. Here it appears that model (c) is the best structural model, a conclusion supported by the finding of scattering in the region of 300 meV, the region where a single bond between hydrogen and sulphur would be expected to have its stretching vibration. Thus, in this particular case, the intensity analysis shows that two related materials, with similar energies of the peaks in their neutron scattering spectra, have different chemical bonding between their hydrogen atoms and the lattice. This must reflect the differing bond strengths of the metal-hydrogen and sulphur-hydrogen bonds in the two materials.

It was mentioned above that these sulphides are used commercially at high temperatures and pressures, and to take advantage of the high penetrating power of neutrons, a cell has recently been constructed [22] to enable neutron scattering measurements to be made under such conditions. By combining inelastic with elastic (diffraction) neutron scattering it will be possible to explore the microstructural origins of the enthalpy transitions [23] and discontinuous changes in rate constants [24] that have been observed for these materials at high pressures of hydrogen. In addition there is also the prospect of exploring whether at high concentrations of sorbed hydrogen there is evidence in the vibration spectra of hydrogen-hydrogen interactions similar to those that have been observed for the hydrogen-palladium system [25, 26].

7.2.2 Butane Adsorption Upon Graphite

In this example, simulated neutron spectra have been compared with experimental data, to obtain information on the relative orientation of an adsorbed molecule to a surface [27]. Butane was chosen for this purpose, primarily because its low energy intramolecular excitations were already well understood from neutron-scattering experiments on the bulk solid and liquid [28]. Three models were chosen to represent the system and to provide a basis for the calculations.

In model (a), the plane of the molecular skeleton of the butane was perpendicular to the basal planes of the graphite with bonding through the hydrogen atoms; in (b) and (c) it was parallel to the same planes, with bonding through the hydrogen and carbon atoms, respectively. Two force constants were introduced as variables, one between the surface and the two CH_2 groups, K_2, and another between the surface and the CH_3 groups, K_3. In model (b), K_2 joined the surface to the hydrogen atoms. In model (c) it involved the carbon atoms of the CH_2 groups. Predictions based on the three models were compared with the experimental data, as shown in Fig.6, and only model (b) was able to provide a good description of the experimental data. Underlying the scattering from the intramolecular vibrations, was additional scattering arising from the intermolecular vibrations. This requires much more detailed structural information for accurate simulation, so that simple spectral analyses will only be possible for molecules where the inter- and intramolecular vibrations are well separated; a circumstance which does not appear to be the case for ethane.

The figure shows that both the frequencies and the relative intensities of the surface vibrations depend strongly on the chosen model for the bonding. These results add emphasis to the contention that even for moderately sized

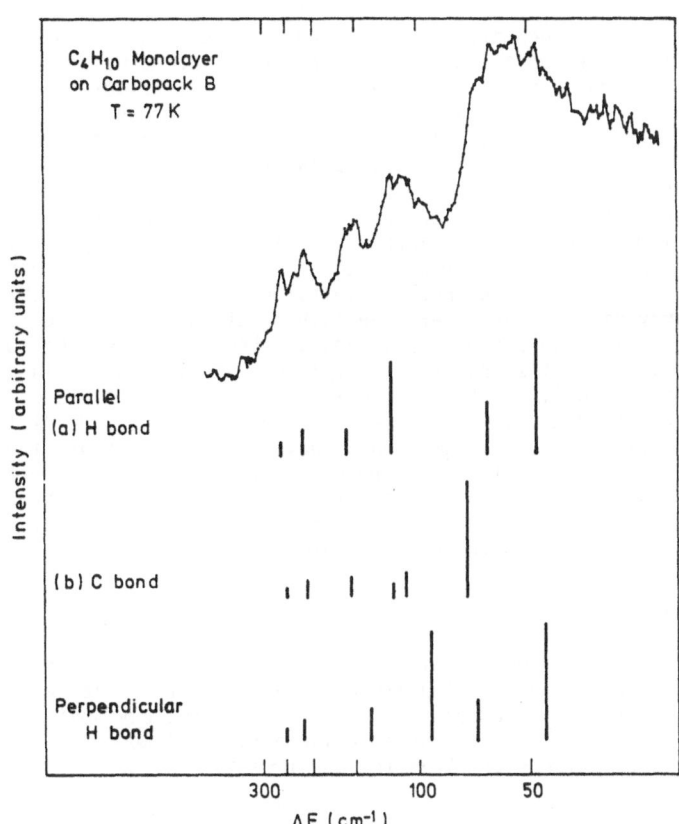

Fig. 6. A comparison of experimental and theoretical scattering data for butane adsorbed on graphite

molecules it is necessary to undertake intensity calculations in order to identify properly the origins of the individual excitations.

7.2.3 C_6H_6 Adsorbed on Raney-Nickel

In this example, analysis of the intensity information provided evidence for the existence of molecular dissociation at a surface. The system chosen was benzene adsorbed upon the surface of Raney nickel at 300°C [29], and force constants, obtained from an analysis of the inelastic neutron scattering from chromium benzene tricarbonyl, were used to predict the spectra [30]. It was found that there were significant discrepancies between measurement and prediction in the region of $1000 \, cm^{-1}$, which could only be reconciled by adding to the predicted spectra a contribution due to the vibrations of nickel-hydrogen bonds. Good agreement between theory and experiment was only obtained if it was assumed that 10% of all the benzene molecules had dissociated.

7.2.4 Collective Excitations

The final applications to be discussed here, are those where use is made of the wide range of momentum transfers accessible in a neutron experiment so that collective excitations of adsorbed molecules can be detected and observed.

In one experiment use has been made of the strong coherent scattering from ^{36}Ar. A sample of oriented graphite covered with argon was prepared and scattering measurements were recorded as a function of momentum transfer parallel to the graphite basal planes. Collective excitations were observed and the dispersion relations were determined sufficiently well to obtain the Ar-Ar force constants in the two dimensional adsorbate layers [31]. It was found that these were essentially similar to the Ar-Ar force constants in bulk Ar. It is hoped that with either better oriented high surface area materials or with higher neutron fluxes further experiments of this type will be performed in the future. This experiment has been paralleled by recent interpretations of the variation in the inelastic incoherent scattering from hydrogen adsorbed at the surface of Raney nickel, and determined as a function of coverage [32]. In an incoherent scattering experiment the intensity observed is related to the density of vibrational states, so that it still contains, although in a less readily accessible manner, information on the dispersion relations of the system. Side bands of 80 and 140 meV were observed, on either side of the central peak in the scattering at 113 meV, which have been assigned previously to the perpendicular excitations of the adsorbed atoms. The side bands were interpreted as peaks in the density of states of the vibrations of the hydrogen atoms parallel to the metal surface; the splitting arising because of the interaction force constant between the adsorbed atoms.

7.3 Future Considerations

The inelastic-scattering instruments that have been used for most of the experiments in this chapter, are high-intensity machines appropriate for examining small quantities of surface-adsorbed material. Unfortunately neither of these machines was designed to have access to a large range of momentum transfers at a particular energy transfer. This follows from the relationships

$$Q^2 = K_0^2 + K_1^2 - 2K_0K_1 \cos\theta \qquad (10)$$

and

$$\hbar\omega = \hbar/2m\left(K_0^2 - K_1^2\right)$$

and the instrumental requirement that K_0 and K_1 must be small for the time-of-flight and beryllium-filter spectrometer, respectively. Consequently for high energy transfers Q is large and relatively independent of scattering angle. Scattering from harmonics and multiple scattering are high, since their intensities are proportional to higher orders of Q^2; whilst at the same time it is impossible to separate these components from the single scattering due to fundamentals, by resolving the total intensity into different contributions of different momentum transfer dependence.

An illustration of the problems that this brings can be seen in the scattering from hydrogen adsorbed by Raney nickel. At energy transfers near to 250 meV it was thought that scattering occurs due to the fundamental vibration of a linear hydrogen nickel bond together with the first harmonics of vibrations due to multiple hydrogen nickel bonds [19]. Since the Q dependence of the scattering could not be determined, a resolution of the relative contributions of bridged and single bonded hydrogen atoms was not possible. Instruments such as triple-axis spectrometers [15,14], which do have provision for varying Q have counting rates which are too low to be of present use in surface spectroscopy.

A new approach to this problem, however, is provided by an instrument which has recently been commissioned on the linear accelerator source at Harwell, the inelastic rotor spectrometer [33]. This instrument exploits the principle that scattering at low momentum transfers can be obtained as long as the conditions:

$$K_0 \simeq K_f$$

or

$$E_0, E_f \gg \Delta E$$

are fulfilled. Under these conditions $Q^2 \to 2K_0^2(1-\cos\theta)$ and $Q^2_{\theta \to 0} \to 0$.

The instrument makes use of pulses of undermoderated high-incident-energy neutrons in the $0.25 \to 1.0$ eV range. The neutron pulses are monochromated by a chopper, and energy analysed, after scattering, by time-of-flight techniques. This facility allows the observation of excitations at 125 meV, for example, using 500 meV incident energy neutrons with momentum transfers down to 15% of those possible with a beryllium-filter spectrometer.

This instrument, and others like it under construction elsewhere, will produce much interesting new information. This will help us to understand existing problems of interpretation, and provide new results especially in the realm of two-dimensional vibrational densities of states.

References

1. H. Taub, L. Passell, J.K. Kjems, K. Carneiro, J.P. McTague, J.G. Dash: Phys. Rev. Lett. *37*, 1695 (1976)
2. W. Marshall, S.W. Lovesey: *Theory of Thermal Neutron Scattering* (Oxford University Press, Oxford 1971)
3. J.A. Janik, A. Kowalska: In *Thermal Neutron Scattering*, ed. by P.A. Egelstaff (Academic Press, London, New York 1965)
4. G.E. Bacon: *Neutron Diffraction* (Oxford University Press, Oxford 1975)
5. C.J. Wright: J. Chem. Soc. Faraday II *73*, 1497 (1977)
6. C. Riekel, H.G. Reznik, R. Schollhorn, C.J. Wright: J. Chem. Phys. *70*, 5203 (1979)
7. C.J. Wright, D. Fraser, R.B. Moyes, P.B. Wells: A.E.R.E. Harwell Rpt. MPD/NBS/162
8. R. Stockmeyer, H.M. Conrad, A.J. Renouprez, P. Fouilloux: Surf. Sci. *49*, 549 (1975)
9. J. Howard, T.C. Waddington, C.J. Wright: J. Chem. Phys. *64*, 3897 (1976)
10. C.J. Wright, C. Sampson, D. Fraser, R.B. Moyes, P.B. Wells, C. Riekel: J. Chem. Soc. Faraday I *76*, 1585 (1980)
11. R.M. Brugger: In *Thermal Neutron Scattering*, ed. by P.A. Egelstaff (Academic Press, London, New York 1965)
12. P.H. Gamlen, N.F. Hall, A.D. Taylor: A.E.R.E. Harwell Rpt. USS/P29 (1974)
13. G.C. Stirling: In *Chemical Applications of Thermal Neutron Scattering*, ed. by B.T.M. Willis (Oxford University Press, Oxford 1973)
14. P.K. Iyengar: In *Thermal Neutron Scattering*, ed. by P.A. Egelstaff (Academic Press, London, New York 1965)

15. J. Howard, T.C. Waddington, C.J. Wright: J. Chem. Soc. Faraday II *73*, 1768 (1977)
16. J. Howard, T.C. Waddington, C.J. Wright: Chem. Phys. Lett. *56*, 258 (1978)
17. R. Stockmeyer, H.M. Conrad, A.J. Renouprez, P. Fouilloux: Surf. Sci. *49*, 549 (1975)
18. A.J. Renouprez, P. Fouilloux, G. Coudurier, D. Tochetti, R. Stockmeyer: J. Chem. Soc. Faraday I *73*, 1 (1977)
19. J. Howard, T.C. Waddington, C.J. Wright: Neutron Inelastic Scattering I.A.E.A. Vienna Vol. *2*, 499 (1978)
20. M.W. Newbery, T. Rayment, M.V. Smalley, R.K. Thomas, J.W. White: Chem. Phys. Lett. *59*, 461 (1978)
21. C.J. Wright, C. Sampson: To be published
22. C.J. Wright, W.A. England, D. Price: To be published
23. D.H. Broderick, G.C.A. Schuit, B.C. Gates: J. Cat. *54*, 94 (1978)
24. D.G. Hunt, D.K. Ross: J. Less-Common Met. *49*, 169 (1976)
25. T. Springer: In *Hydrogen in Metals I*, ed. by G. Alefeld and J. Völkl, Topics in Applied Physics, Vol.28 (Springer, Berlin, Heidelberg, New York 1978)
26. J.M. Rowe, J.J. Rush, H.G. Smith, M. Mostoller, H.E. Flotow: Phys. Rev. Lett. *32*, 1297 (1974)
27. H. Taub, H.R. Danner, Y.P. Sharma, H.L. McMurry, R.M. Brugger: Phys. Rev. Lett. *39*, 215 (1977)
28. K.W. Logan, H.R. Danner, J.D. Gault, H. Kim: J. Chem. Phys. *59*, 2305 (1973)
29. H. Jobic, J. Tomkinson, J.P. Candy, P. Fouilloux, A.J. Renouprez: Surf. Sci. *95*, 496 (1980)
30. H. Jobic, J. Tomkinson, A. Renouprez: Mol. Phys. *39*, 989 (1980)
31. H. Taub, K. Cameiro, J.K. Kjems, L. Passell, J.P. McTague: Phys. Rev. B *16*, 4551 (1977)
32. R. Stockmeyer, H. Stortnik, I. Natkanic, J. Mayer: Ber. Buwsenges. Phys. Chem. *84*, 79 (1980)
33. B.C. Boland, D.E.R. Mildner, G.C. Stirling, L.J. Bunce, R.N. Sinclair, C.G. Windsor: Nucl. Instr. Meth. *154*, 349 (1978)

8. Reflection Absorption Infrared Spectroscopy: Application to Carbon Monoxide on Copper

P. Hollins and J. Pritchard

With 10 Figures

8.1 Background

In this paper we discuss the application of reflection-absorption infrared spectroscopy (RAIRS) to the study of adsorbates on metal single crystal planes and the use of RAIRS for the interpretation of the transmission infrared spectra obtained from supported metals and catalysts. The spectra of carbon monoxide on copper will be emphasized, but the much wider scope of RAIRS will also be indicated.

The usefulness of infrared spectroscopy as a surface sensitive method for the study of adsorption on supported metals was originally demonstrated by EISCHENS and his co-workers [1-3]. In the course of that early work it appeared that the ν_{CO} bands in the transmission spectra of chemisorbed CO on a variety of metals could be interpreted by analogy with the spectra of molecular carbonyls in terms of linear and bridging groups. Linear absorption implies bonding to one metal atom at an on-top site, while a bridging group is adsorbed at a site of higher coordination number and is involved in bonding to two or more metal atoms. In a recent extensive review SHEPPARD and NGUYEN [4] consider this distinction to remain essentially valid, but we shall see that there is a problem with the spectra of dense CO adlayers where only one band may occur despite the evidence from LEED for out-of-registry adsorption.

Of particular interest was the observation by EISCHENS et al. [2] that with increasing coverage the spectrum of CO adsorbed on silica-supported palladium showed a reversible development of structure in the broad band attributed to bridging CO groups, whereas the strong band of linear CO on platinum only shifted to higher frequencies but did not alter in shape. The latter behaviour is connected with intermolecular dipole coupling [2,5], but it was suggested that the former could be related to adsorption on a patchy surface composed of facets with the major crystallographic orientations. Support for this idea was provided by BADDOUR et al. [6] who found that the shape of the envelope of the bands changed as the supported palladium was progressively broken in as a catalyst for CO oxidation. Such changes could be caused by structural changes in the surfaces of the metal particles under catalytic conditions, leading to the development of preferred facets. However, PALAZOV et al. [7] concluded that the changes of the spectra with coverage were caused more by intermolecular interactions than by the sequential population of different facets. An alternative explanation for the break-in effect, proposed by ENGEL and ERTL [8], is the oxidative removal of impurities such as carbon and sulphur from the original palladium surface.

The CO-palladium system illustrates the difficulty presented by the lack of definition of supported metal surfaces when deducing adsorbate behaviour from transmission infrared spectra. On the other hand it also indicates the potential that such spectra may offer for the characterization of real catalyst surfaces if reference spectra from surfaces of known structure and composition can be established. Another illustration of the connection between spectra and surface structure occurs in the case of nitrogen on nickel. The first infrared spectra of dinitrogen acting as a ligand were obtained by EISCHENS and JACKNOW [9]. A strong band at 2200 cm^{-1} was formed for nitrogen adsorbed on silica-supported nickel. It was shown subsequently [10] that the intensity of the band depended strongly on the size of the supported nickel particles and that specific sites of five-fold coordination (B_5) were probably necessary for nitrogen to be adsorbed in this infrared active form. The kinds of site suggested would be abundant on planes such as {110}, {311}, and stepped surfaces, but not on the smooth low index faces. These examples clearly show the need for spectra to be obtained from well defined single crystal surfaces. Transmission spectroscopy cannot be used with bulk metal, nor even with epitaxial metal films which could alternatively provide well-defined surfaces. Reflection methods must be used.

8.2 Physical Factors in Infrared Reflection at Metal Surfaces

Although the surface structure and composition of a supported metal is often poorly characterized its specific surface area is very large. In an infrared reflection experiment the beam encounters at most one monolayer of chemisorbed molecules in each specular reflection. In a transmission experiment with a typical pressed disc sample the equivalent of 10^3 monolayers may be penetrated. The sensitivity of the transmission experiment is therefore high and a large variety of adsorbates may be studied without too much difficulty except in wavelength ranges where the support is opaque. Sensitivity is the major practical problem in reflection experiments and in this section we briefly consider the factors that are of immediate relevance to the experimental approaches to be described later. The importance of the angle of incidence and the polarization of the beam were stressed in the pioneering studies of Blodgett films on metal mirrors by FRANCIS and ELLISON [11]. The theoretical basis was developed further by GREENLER [12], providing the stimulus for further experimental work. More general accounts, both qualitative and quantitative, of specular infrared reflection spectroscopy of adsorbed layers on metals are available elsewhere [13-17].

A beam incident at angle θ to the surface normal may be resolved into s-polarized and p-polarized components (Fig. 1) in which the electric vector oscillation is respectively perpendicular or parallel to the plane of incidence. The electric vectors of both the incident and reflected s-polarized beams are tangential to the metal surface for all values of θ, but as θ increases the vectors of the p-polarized beams, initially tangential too, acquire increasing components normal to the surface. The phase change on reflection at a clean metal surface is always near 180° for the s-component and the amplitude reflection coefficient is close to unity. Thus the resultant of the incident and reflected vectors is practically zero at the surface, and the interaction of s-polarized radiation with adsorbed dipoles is always insignificant. Because of the long wavelengths in the infrared region this conclusion applies not only to monolayers but to quite thick films. A beautiful illustration is provided by POLING's study of thick copper oxalate films on copper [18] in which no sign of the absorption spectrum is seen in reflected s-polarized radiation from films of 25 nm thickness.

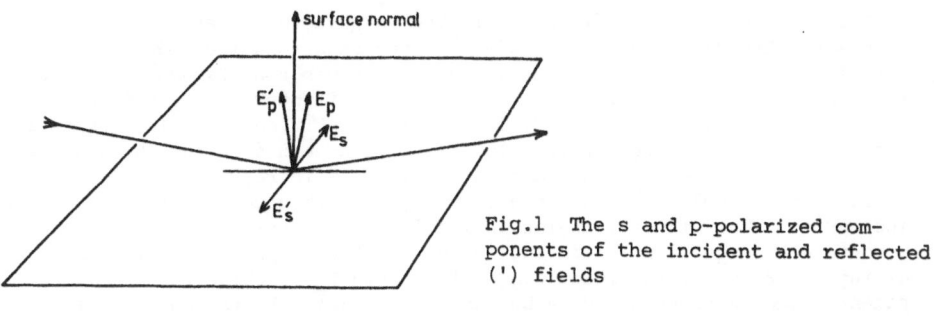

Fig.1 The s and p-polarized components of the incident and reflected (') fields

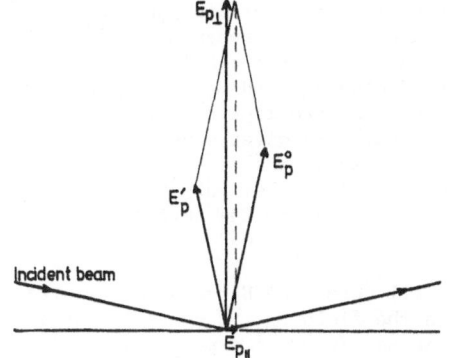

Fig.2 Resultant normal and tangential components of the surface field of p-polarized radiation

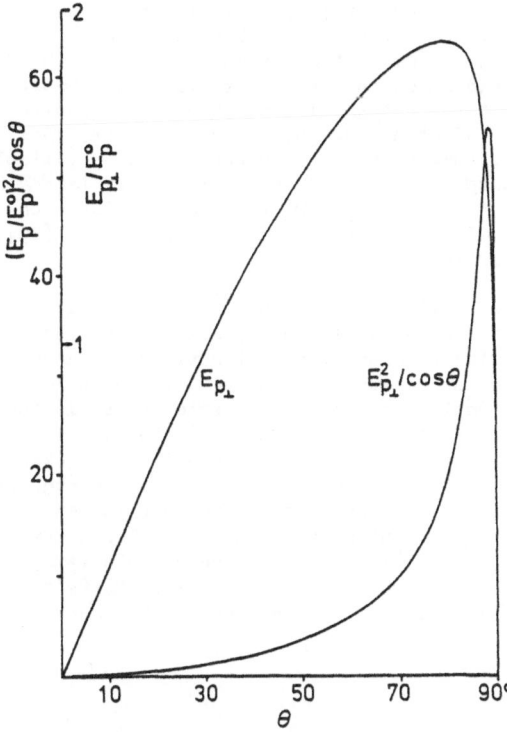

Fig.3 Dependence of E_p and $E_p^2/\cos\theta$ on angle of incidence for highly reflecting metal in near infrared ($n = 3 - 30i$)

The behaviour of p-polarized radiation is quite different. An incident beam with electric vector amplitude E_p^o gives rise to a field $E_p^o\sin\Theta\sin\phi$ normal to the reflecting surface and a field $E_p^o\cos\Theta\sin\phi$ tangential to the surface, where ϕ is an arbitrary phase. The reflected beam yields normal and tangential components $E_p^o r_p\sin\Theta\sin(\phi+\delta)$ and $-E_p^o r_p\cos\Theta\sin(\phi+\delta)$ respectively, where r_p is the amplitude reflection coefficient and δ is the phase change on reflection. Both r_p and δ depend on Θ. As Θ increases δ remains small until at high angles of incidence it changes rapidly towards 180° at grazing incidence. Thus at small angles of incidence the maximum tangential components are large, similar in magnitude ($\sim E_p^o$) but opposed in direction, giving a very small resultant tangential field $E_{p\parallel}$, while the normal components combine constructively but are very small. As the angle of incidence increases, the normal components increase at the expense of the tangential components and the resultant field $E_{p\perp}$ normal to the surface tends to towards $2E_p^o$ before the increasing phase difference eventually causes mutual cancellation. $E_{p\perp}$ therefore passes through a broad maximum value before falling to zero at grazing incidence. The resultant fields $E_{p\perp}$ and $E_{p\parallel}$ (Fig. 2) may be regarded as the major and minor semi-axes of an elliptical standing wave [12]. Unlike the s-component p-polarized radiation gives significant fields at metal surfaces, but only normal to the surface. The metal screens tangential fields and provides an effective selection rule: only those vibrational modes that yield a dipole oscillation normal to the surface are observable in reflection-absorption.

The intensity of the absorption band will depend on $E_{p\perp}^2$ and on the area (number of adsorbed molecules) over which the field $E_{p\perp}$ is exerted. As the area intercepted by a parallel beam varies as $1/\cos\Theta$ we expect qualitatively that the intensity of an actual band should depend on Θ as $E_{p\perp}^2/\cos\Theta$ (Fig. 3). When allowance is made for the modification of the surface field by a thin film this agrees well with quantitative computations [12, 16]. The function $E_{p\perp}^2/\cos\Theta$ peaks sharply at large Θ (at 88° for highly reflecting metals, e.g. Cu, and at only slightly smaller angles for most other metals) and at the optimum angle for absorption of p-radiation in a single reflection is more than an order of magnitude greater than in transmission through the equivalent layer isolated from the metal and at normal incidence. However, this advantage is far from sufficient to compensate for the surface area available in supported metals. The peak absorption from a complete monolayer is small, e.g. 0.25% for the 2210 cm^{-1} band of C_2D_6 on Cu(110) [19], except for such very strongly absorbing species as CO when it may reach as much as 12% [20]. Signal to noise must be carefully optimized [21].

Multiple reflections were used in early work to enhance the relative intensity of the absorption bands. Although the intensity relative to the background continuously increases with the number of reflections the absolute absorbed intensity eventually decreases because of the overall energy loss due to the limited reflectivity of real metals. When detector noise is the factor that ultimately limits the sensitivity it is important to maximize the absolute absorbed intensity, a condition that is reached for weak absorption bands when the number of reflections is such as to reduce the background to e^{-1} or 37% of the incident intensity. Unfortunately, the optimum angle for reflection-absorption is close to the angle at which the reflectivity for p-polarised radiation passes through a minimum (the pseudo-polarizing angle), so that for most metals the absolute absorbed intensity starts to diminish after very few reflections. GREENLER [22] has shown that one reflection is usually sufficient to give at least 70% of the maximum possible absorption signal. In practice it is much simpler to employ a single reflection in single crystal studies and multiple reflections have been abandoned even for the highly reflecting group 1B metals where a real gain in sensitivity is offered. Multiple reflections would be more advantageous if the dominant source of noise were the beam intensity itself. For example, if radiation shot noise were limiting, the optimum number of reflections would be doubled [13].

8.3 Experimental Aspects

The application of RAIR spectroscopy to clean single crystals or evaporated films requires windows that are compatible with ultrahigh vacuum requirements. Ideally the windows should be transparent in the visible region to facilitate optical alignment, of low refractive index to minimize reflection losses, with a wide infrared transmission range, large enough in diameter to accommodate the converging or diverging beams that are necessary with thermal radiation sources, and capable of being sealed to stainless steel flanges for the convenient combination of RAIRS with other surface techniques, such as LEED, AES, etc. [23]. A variety of special window materials and sealing methods have been used [24-26], but for general use alkali halides (NaCl, KBr, etc.) are most satisfactory and they can be attached to flanges with viton O-rings in a bakable differentially pumped configuration that provides excellent ultrahigh vacuum performance [27].

The main experimental problem is that of measuring much lower levels of absorption ($\sim$ 0.1%) than are normal in infrared spectroscopy [16]. At high angles of incidence even large crystals present a slit-shaped aspect. Alignment and focussing of the convergent incident beam is critical, and mechanical instability of the crystal position could cause energy fluctuations larger than the absorption signal. The larger areas available with ribbons or with evaporated films ease this problem. TOMPKINS and ALLARA [28] have described an ingenious optical arrangement of mirror and retro-mirror that facilitate optimization of the alignment for single reflections from evaporated films. In early experiments a single-beam grating spectrometer was used, the spectrum being measured as the difference between spectral scans of the initial clean surface and of the surface with adsorbate. Such experiments are particularly susceptible to low frequency noise and slow drifts in addition to inherent detector noise. The high-angle single-reflection geometry used with single crystals makes it difficult to adapt conventional double-beam spectrometers to operate with truly equivalent paths, but double-beam techniques have been developed to give noise levels as low as 0.01% over limited spectral ranges [29]. Very good performance can be achieved in a single-beam system by either wavelength modulation [19,30] or by polarization modulation [23,31-33].

In wavelength modulation the first derivative of the intensity spectrum is recorded. This discriminates against the large background of unabsorbed radiation, and subsequent integration of the derivative spectrum greatly improves the signal to noise ratio. Polarization modulation exploits the different interactions of s and p-polarized radiation with surface films. In effect the s-component acts as a reference beam in a double-beam system wherein the two beams (s and p) follow identical paths. Polarization modulation may be employed in a direct intensity-comparison mode [23] or as the ellipsometric spectroscopy developed very effectively by DIGNAM and his co-workers [32-35]. An important advantage of polarization modulation is its ability to discriminate between surface absorption, where s and p-components behave differently, from gas phase absorption where the interaction is the same. One can then take full advantage of the capability of infrared spectroscopy, as distinct from electron energy loss spectroscopy, of operating in the presence of ambient gases. This advantage was already apparent in early work on the reversible room-temperature adsorption of CO on copper films; the greater power of polarization modulation has been clearly demonstrated by OVEREND [36,37] for CO adsorption on platinum at up to 0.25 atm using a simple wedge-filter spectrometer [21], a result which indicates the potential of infrared methods for in-situ studies of catalysis on single crystals, and which has been extended to a study of NO adsorption at 50 Torr [38].

8.4 Some Applications of RAIRS

The first chemisorption system to be studied in detail by RAIRS, primarily for reasons of experimental convenience, was that of CO on polycrystalline copper films [39,40]. The adsorption is reversible at room temperature and the linearly bound CO gives a single intense ν_{CO} band at about 2100 cm^{-1}. Such studies of CO chemisorption have been extended to a large number of metals, as evaporated films, ribbons, or single crystals [16]. The intensity of the CO bands is certainly an experimental advantage, but RAIRS is by no means limited to this adsorbate. Examples of applications to hydrogen, nitrogen, nitric oxide, ethane, ethylene, acetylene and formic acid have been reviewed elsehwere [16].

Before returning to the topic of CO adsorption on copper some results of particular relevance to the surface structure theme of the Introduction should be mentioned. Firstly, the idea based on particle size effects that the infrared active dinitrogen species adsorbed on supported nickel are associated with B$_5$ sites led to the prediction that Ni(110) should give such species [10]. This is confirmed by a recent single crystal study [41]: dinitrogen is adsorbed with a heat of 35 kJ mol^{-1} and gives an infrared band at 2194 cm^{-1}. Secondly, BRADSHAW and HOFFMANN [31] have reported spectra of CO on (111), (100) and (210) single crystal faces of palladium that can be compared with the spectra of CO on supported palladium. Together with LEED evidence their results confirm the attribution of the low frequency bands to bridging CO groups. In the half-coverage $(2\sqrt{2} \times \sqrt{2})$-45° overlayer structure of CO on Pd(100) all the molecules occupy two-fold bridging sites [42] and the infrared band appears at 1949 cm^{-1}. At low coverage the band appears at 1895 cm^{-1}, so there is a large frequency shift with coverage. The shift is similar to the upward frequency shift of linearly bound CO on supported platinum [2], platinum films [43], and Pt{111} [29,44], which has been shown to be caused by coupling interactions by the isotopic decoupling experiments of CROSSLEY and KING [45]. An even larger shift occurs on Pd(111), from 1823 cm^{-1} at low coverage to 1936 cm^{-1} at half coverage, but in addition to coupling interactions the adsorption sites are believed to change from threefold to two-fold bridges. In the light of the single crystal results SHEPPARD and NGUYEN [4] consider that the effect of "break-in" on the CO spectra from supported palladium can be attributed to the growth of {111} and particularly {100} facets.

8.5 Spectra of CO on Copper

The spectra of CO adsorbed on copper have been discussed in detail elsewhere [46]. We summarise the main aspects here before discussing the significance of dipole coupling effects in these spectra and for the comparison of spectra from single crystals with those from supported copper. Spectra of CO on evaporated copper films show essentially a single intense band. The peak frequency has been found to be remarkably reproducible in different investigations, lying in the range 2102 - 2107 cm^{-1} and showing negligible shift with increasing coverage. Transmission spectra of CO on silica or alumina-supported copper are very similar, but the range of peak frequencies can extend to 2146 cm^{-1} depending on the preparation and pretreatment of the catalyst [47]. Compared with the spectra of CO on metals such as palladium and platinum the lack of frequency shift with coverage is very striking.

Surprisingly, the low index single crystal faces give bands at appreciably lower frequencies, at 2080 cm^{-1} on Cu(100) and 2076 cm^{-1} on Cu(111), whereas the higher index surfaces so far studied - (110), (311), (211), (755) - give bands in the range 2096 - 2110 cm^{-1} in reasonable agreement with the results from polycrystalline films deposited on glass and from silica supported copper. A diagrammatic comparison of the bands from the single crystals with that from a polycrystalline film on glass is shown in Fig.4.

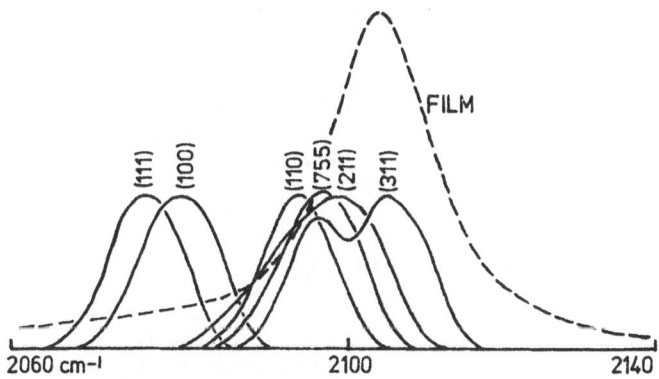

Fig.4 Comparison of single crystal band position with spectrum of CO on polycrystalline copper film

These results are believed to indicate the virtual absence of low index surfaces in ordinary polycrystalline film surfaces or on supported copper crystallites. This conclusion receives support from the effect of choosing a substrate or support that is known to give epitaxial growth of the metal with low index orientations. With magnesium oxide the CO bands appear at about 2080 cm^{-1} [48]. The lack of smooth low index faces in the surfaces of thick films deposited on amorphous substrates is probably a consequence of the very small anisotropy of the surface energy of clean copper [49]. Electron microscopy of supported copper shows the development of spherical particles [47].

The comparison of the spectra of CO on single crystals with those from films and supported copper is greatly eased by the insensitivity of the band positions to surface coverage. Consequently, spectra from supported copper at room temperature, where the coverages of individual facets are pressure dependent and will vary from one to another, can be compared with those from single crystals at low temperatures and pressures. Shifts with coverage do occur on the single crystals, but in the pre-compression range they are very small and may be of either sign. The bands are comparable in intensity with those of CO on platinum and, in general, dipole coupling is expected to cause an upward frequency shift. However, we shall see that it is very difficult to predict the magnitude of that shift; it must be determined by isotopic decoupling. Chemical effects, such as competition for the electrons involved in synergetic σ and π-bonding, may cause shifts of either sign [46]. It is important to distinguish these contributions to the overall behaviour and to assess the possible influence of facet size on the frequency shifts resulting from simultaneous chemical influences and long-range dipole coupling.

8.6 Coupling Effects in Vibrational Spectra

Coupling effects, which give rise to frequency shifts and intensity changes in vibrational spectra, have already been extensively studied in the solid state. A detailed discussion of their significance in molecular crystals has been given by DECIUS and HEXTER [50]. Coupling occurs when the expression for the vibrational potential energy contains a term of the form $\Sigma a; a_j Q_i Q_j$ involving the products of the normal coordinates Q_i, Q_j of two molecules. In the solid state such terms can arise from either a long-range dipolar coupling [51] or a much shorter range van der Waals type interaction [52].

The first attempt to use a coupling model to account quantitatively for changes in the spectrum of an adsorbed species was made by HAMMAKER, FRANCIS and EISCHENS (HFE) in their study of CO adsorption on supported Pt [5]. By treating the adlayer as an array of N parallel oscillating point dipoles they derived the following expression for the potential energy V

$$2V = \sum_{i=1}^{N} 4\pi^2 c^2 \omega_{oi}^2 Q_i^2 + 2 \sum_{i>j=1}^{N} R_{ij}^{-3}(\partial\mu/\partial Q_i)(\partial\mu/\partial Q_j)Q_i Q_j \tag{1}$$

where R_{ij} is the distance between molecules i and j, ω_o is the frequency of an isolated singleton molecule, and $\partial\mu/\partial Q$ is the derivative of the dipole moment with respect to the normal coordinate.

For an isotopically pure adsorbate this expression yields an exact solution for the frequency ω of the only infrared-active mode in which all the molecules vibrate in-phase:

$$\omega^2 = \omega_o^2 + (\partial\mu/\partial Q)^2 \sum_{j=2}^{N} R_{ij}^{-3}/4\pi^2 c^2 \tag{2}$$

It is convenient to replace $\partial\mu/\partial Q$ by the vibrational polarizability α_v to which it is related by

$$\alpha_v = (\partial\mu/\partial Q)^2/4\pi^2 c^2 \omega_o^2 \tag{3}$$

Thus

$$(\omega/\omega_o)^2 = 1 + \alpha_v T \tag{4}$$

where T is the dipole sum ΣR_{ij}^{-3}. Since T increases with coverage the frequency also increases.

132

HFE also derived approximate expressions for the frequencies of the modes which arise when a molecule of one isotopic species is surrounded by N-1 identical molecules of a different species. There are two infrared active modes: one at frequency ω_h in which the labelled molecule vibrates in-phase with its environment, and one at a lower frequency ω_l in which it vibrates 180° out-of-phase. These frequencies are related to the frequency, ω_2, of the labelled singleton and ω_l^{1}, the frequency of the surrounding 2-D lattice in the absence of the labelled molecule, by

$$(\omega_l/\omega_2)^2 = 1 - \alpha_v^2 \Sigma R_{ij}^{-6} \omega_l^{1^2} / (\omega_l^{1^2} - \omega_2^2) \tag{5}$$

$$(\omega_h/\omega_l^{1})^2 = 1 + \alpha_v^2 \Sigma R_{ij}^{-6} \omega_2^2 / (\omega_l^{1^2} - \omega_2^2) \tag{6}$$

The frequency shifts implied by these formulas are very much less than that for the single isotope case, as one might anticipate from the fact that coupling is strongest between identical oscillators. Hence, in a dilute isotopic mixture the minor component is effectively decoupled from its environment. As HFE pointed out, coupling interactions also bring about a pronounced transfer of intensity into the high frequency band in a mixed isotope spectrum.

These effects enable one to design experiments which will distinguish coupling shifts - dipolar or otherwise - from other ("chemical") shifts, which involve a change in ω_0. CROSSLEY and KING [45,53] have shown that the 36 cm^{-1} shift observed for CO on Pt(111) as the coverage increases can be reproduced at a constant coverage by varying the isotopic composition from a dilute mixture of ^{12}CO in ^{13}CO to pure ^{12}CO demonstrating that the whole of the shift arises from coupling interactions rather than from chemical effects, which would be independent of isotopic composition.

As described above, the shifts observed for CO adsorbed on copper are relatively small, and may occur in either direction, depending on the particular crystal face and the coverage range being considered. Three hypotheses may be introduced to account for this pronounced difference from the transition metals:

i) Coupling interactions on copper may be very weak.

ii) The monolayer may develop by an island growth mechanism, so that even at very low coverages no singletons occur. Such behaviour has been postulated [54] for the adsorption of CO on Pt(111) at 80 K, where the IR band was observed to grow at constant frequency. CROSSLEY and KING [53] have also obtained evidence - in this case supported by mixed isotope spectra - for island growth of CO on Pt(100) over a wide range of temperatures.

iii) Large coupling shifts may occur, but these may be balanced by large opposing chemical shifts.

These three possibilities can be readily distinguished by studying the spectra of ^{12}CO - ^{13}CO mixtures over a range of compositions and coverages:

i) Weakly-coupled CO should give rise to spectra showing two bands having intensities approximately proportional to the relative isotopic abundances and with a frequency separation equal to the gas-phase value (47 cm^{-1}) at all coverages.

ii) Islands of strongly-coupled CO molecules should yield spectra demonstrating anomalous intensities and frequency separations even at very low coverages.

iii) If there are two opposing shifts, dilute isotopic mixtures should give a band, due predominantly to the minor species, which is decoupled from its environment and hence exhibits the chemical shift to lower frequency, whereas the major band should show behaviour similar to that of the isotopically pure adsorbate. Intensity ratios should become increasingly anomalous as the coverage rises.

Spectra have been obtained as a function of coverage and isotopic composition on Cu(111) [20], the face giving the most strikingly anomalous behaviour in that the frequency of the single-isotope band moves continuously downwards with increasing coverage. These spectra (Fig.5) demonstrate conclusively that the third hypothesis is correct: a strong coupling shift is opposed by an even stronger chemical shift.

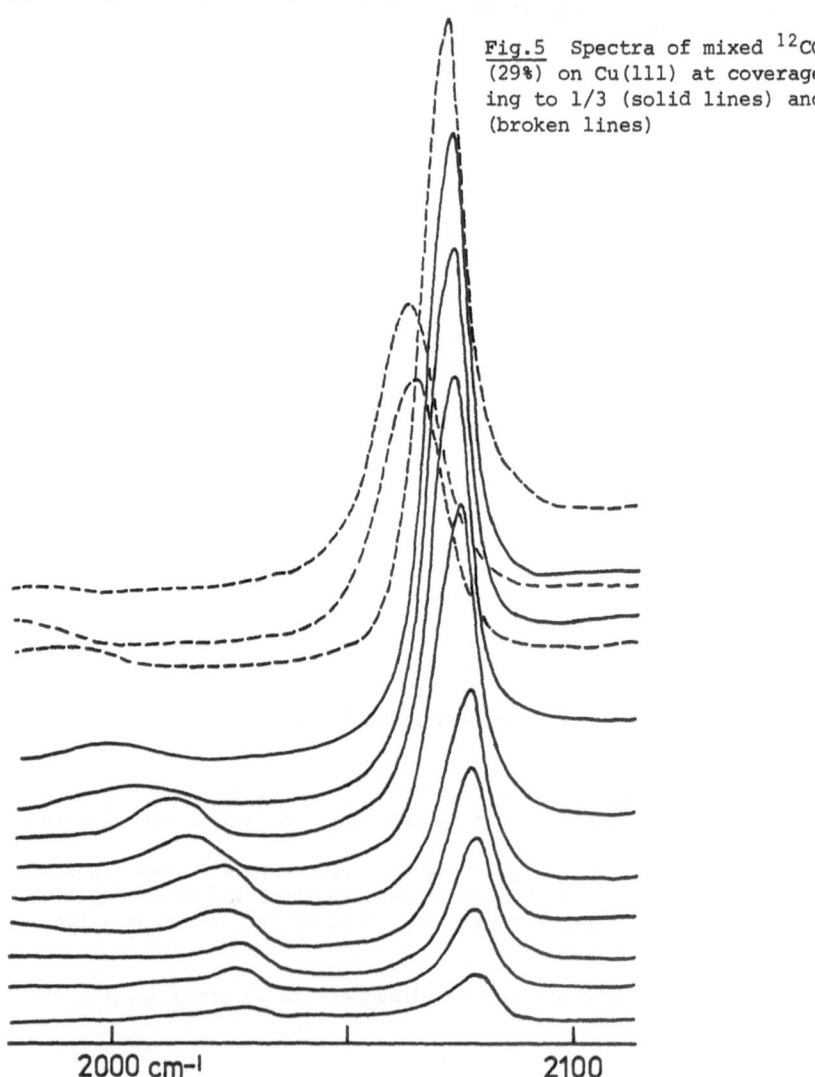

Fig.5 Spectra of mixed ^{12}CO and ^{13}CO (29%) on Cu(111) at coverages increasing to 1/3 (solid lines) and beyond (broken lines)

2000 cm⁻¹   2100

PERSSON and RYBERG [55] have recently carried out a RAIRS study of CO on Cu(100), varying the isotopic composition at constant coverage ($\theta = \frac{1}{2}$). Their results indicate that opposing shifts occur in this system as well, and it seems reasonable to expect similar behaviour on the other faces of copper.

A fortuitous balance of opposing shifts is not however confined to the CO/Cu system. A similar effect was proposed many years ago by VAN HARDEVELD and VAN MONTFOORT to account for the constant frequency of N_2 adsorbed on Ni [56]. Recent work carried out using $^{14}N_2-^{15}N_2$ mixtures has confirmed this for N_2/Ni(110) [57].

8.7 Modified Dipole Coupling Theories

Although the experimental work described above has clearly demonstrated the existence of coupling interactions in adlayers, there remains some uncertainty about their theoretical description, and, in particular, about the extent to which dipole coupling alone can account for the observed effects.

If one substitutes the value of α_v for the gas-phase molecule into the equations developed by HFE, one concludes that the observed frequency shifts are far greater than those predicted by the theory. This can be remedied by assuming a very much larger value for α_v, as CROSSLEY and KING [45] did originally. Increased values of α_v (or $\partial\mu/\partial Q$) are not in themselves unreasonable, and very large values have been found in certain metal carbonyls [58]. However, it can be shown that in the HFE model the band intensity should be proportional to $(\partial\mu/\partial Q)^2$ and, as we have discussed elsewhere [46], the observed intensities are not compatible with the value of $\partial\mu/\partial Q$ deduced from the shift. It therefore seems that the HFE treatment is inadequate.

An obvious deficiency of the model - clearly recognised by the original authors [5] - is that it ignores any influence of the metal surface (other than that of orienting the dipoles). An attempt to devise a more realistic model was made by MAHAN and LUCAS [59], who modified the theory in two ways. Firstly, they assumed a classical model for the metal surface, allowing the reference dipole to interact not only with the other dipoles, but also with their images. Secondly, they considered the screening influence of the adlayer, assigning an electronic polarizability, α_e, to the CO molecule. The combined effect of these two changes was to alter (4) to

$$(\omega/\omega_o)^2 = 1 + \alpha_v S/(1 + \alpha_e S) \tag{7}$$

where S is the sum over all dipoles (other than the reference molecule's) and their images.

MAHAN and LUCAS concluded that even with these modifications the theory was incapable of producing sufficiently large shifts if α_v and α_e were assumed to adopt their gas-phase values. They also found that the shift was relatively insensitive to the distance d between the centre of the adsorbed dipole and the image plane.

SCHEFFLER [60] then modified the theory further by including the reference molecule's self-image in the image dipole sum. If we define S as before and introduce the self-image term explicitly the expression for the shift becomes

$$(\omega/\omega_o)^2 = 1 + \alpha_v (S - 1/4d^3)/[1 + \alpha_e (S - 1/4d^3)] \tag{8}$$

The effect of introducing the self-image is profound. ω_o no longer represents the singleton frequency, but is instead the frequency of a hypothetical molecule isolated not only from other dipoles, but also from its self-image. The singleton frequency, i.e. $\omega(S = 0)$, is given by

$$\omega(o) = \omega_o[1 - \alpha_v/(4d^3 - \alpha_e)]^{\frac{1}{2}} \quad . \tag{9}$$

This falls to zero when

$$4d^3 = \alpha_e + \alpha_v = \alpha_s \tag{10}$$

at which point the positive feedback produced by the self-image field causes the induced polarization to increase without bound. When d is close to the value given by (10) extremely large shifts can be obtained even if α_v and α_e have gas-phase values, and the size of the shift is extremely sensitive to small changes in d, as Fig.6 shows.

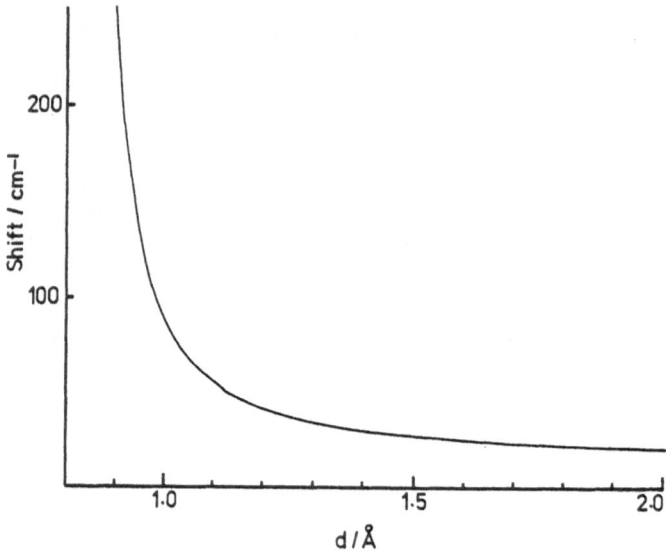

Fig.6 Frequency shift from zero coverage as a function of the distance d for $\overline{Cu}(111)$-CO$(\sqrt{3} \times \sqrt{3})$-$30°$ structure ($\alpha_v = 0.156$ Å^3, $\alpha_e = 2.3$ Å^3, $\omega_o = 2089$ cm^{-1})

Several authors [53,61,62] have demonstrated that their observed shifts are consistent with (8) using gas-phase values for α_v and α_e, but in none of these cases has detailed consideration been given to the variation of the band intensity with coverage. Within the SCHEFFLER model and assuming narrow, constant line-width, the following relationship holds [63]

$$[\ln(I_o/I)]_{peak} \propto \alpha_v\Theta[1 + \alpha_e(S - 1/4d^3)]^{-2} \quad . \tag{11}$$

We have applied SCHEFFLER's theory to the CO/Cu(111) system, for which detailed coverage information is available for $\Theta \leqslant 1/3$ [64]. Although the coverage dependence of the shift can be adequately reproduced using gas-phase

α_v and α_e values, the predicted intensities are very different from those observed in practice. We also find that the value we are constrained to use for d (1.07 Å) differs considerably from the value we would anticipate from LEED data on comparable systems [65,66].

Both of these faults can be remedied by allowing α_v and α_e to assume values which differ from those for gaseous CO. First we estimate d to be 1.63 Å from the LEED data. (As Fig.6 shows the shift is not very sensitive to changes in d for image distances this large, and our conclusions are not very much changed if this value is substantially in error). Values of α_e = 2.3 Å^3 and α_v = 0.156 Å^3 then enable us to fit both the frequency and the intensity data (Fig.7).

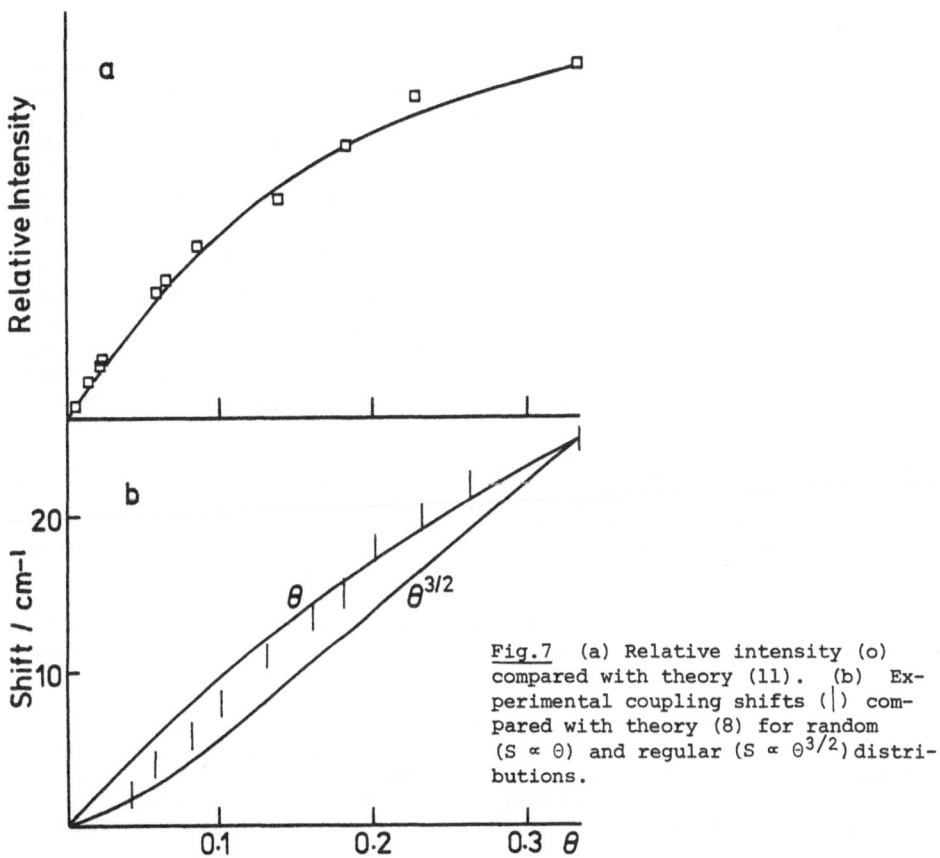

Fig.7 (a) Relative intensity (o) compared with theory (11). (b) Experimental coupling shifts (|) compared with theory (8) for random (S ∝ θ) and regular (S ∝ θ$^{3/2}$) distributions.

The value we find for α_e is close to that for the gas-phase molecule (2.54 Å^3), but the value of α_v is much larger (α_v(gas) = 0.057 Å^3). As mentioned previously, such a figure is not unreasonable when one considers the range found for CO ligands, and a comparable value has been deduced by PERSSON and RYBERG for CO/Cu(100).

Apart from the range 0 < θ ⩽ 1/3 already considered, we have accurate coverage information for CO/Cu(111) at saturation and at one intermediate point [67]. From the spectra obtained at these coverages it is clear that

the dipole coupling model cannot reproduce the experimental results for
$\theta > 1/3$ if the same values of α_e, α_v and d are used. This is not unexpected,
for two distinct reasons. First, there is clear evidence from SP, LEED,
thermodynamic and infrared line-width studies [20]that adsorption beyond
$\theta = 1/3$ leads to pronounced changes in the adlayer, changes which one would
expect to find reflected in the molecular parameters. LEED indicates the
formation of out-of-registry structures which cannot be reconciled with
simple on-top adsorption sites alone, yet there is no sign of bands which
could be ascribed to bridging species [46]. Secondly, it is generally ap-
preciated from solid state studies that dipole coupling theories become
inadequate when the interacting molecules are very close [50], and short
range forces must then be taken into account: at saturation the CO molecules
are only 3.55 Å apart, a distance comparable with their van der Waal's radius
of 3.3 Å.

In the light of these results, how adequate has dipole coupling theory
shown itself to be?   There is as yet no experimental evidence to indicate
that the theory fails under the conditions of low packing density where it
might be expected to prevail, and it has not been necessary to invoke the
"vibrational coupling" introduced in an ad hoc fashion by MOSKOVITS and
HULSE [68] to account for direct through-metal effects. Nonetheless, the
approximations we have used - of classical images and point dipoles - are
extreme, and it seems likely that the inadequacies of these concepts are to
a certain extent camouflaged by the recourse to a large number of free parameters. EFRIMA and METIU [69] have considered the electrodynamics of an adsorbed dipole of finite length, but this necessarily introduces additional
parameters. Since our current understanding of chemisorption is too meagre
to enable us to calculate all these molecular parameters from first principles,
all applications of the theory have been rationalisations of observed results,
and its predictive power is virtually nil.

Although our understanding of the theory of coupling interactions in adlayers is still very limited, their practical importance is considerable.
Much detailed knowledge of adsorption systems can be obtained by comparing
their vibrational spectra with those of free molecules of known structure,
as SHEPPARD has pointed out during this conference, but the effect of coupling interactions in the adsorbed layer must not be ignored. However, even
the simplest theoretical treatment shows how coupling shifts may be determined
experimentally, and their measurement has become important in characterizing
an adsorption system's properties.

An application of dipole-coupling theory which is likely to achieve pro-
minence in the near future is its use to establish whether adlayer formation
proceeds by an island growth mechanism. Fortunately, it seems that the properties predicted for various sizes of island are not too strongly dependent
on the details of the dipole-coupling theory employed, as will be shown in
the final section of this paper.

8.8 Dipole Coupling in Islands

It is now evident that the previously bewildering variety of frequency shifts
observed on the different crystal faces of copper can be explained as a consequence of rather small changes in the relative balance of two large but
opposing effects, dipole coupling and the chemical shift. In deducing from
the spectra shown in Fig.4 that polycrystalline films consist predominantly
of high-index facets, we have made the implicit assumption that this balance
is not altered by the difference in size between the small facets of the film

and the much larger samples used for single crystal work. However, dipole coupling is known to be a long-range effect whereas the chemical shift may be limited to a few lattice spacings. Consequently, the truncation of the dipole sum which occurs when the substrate is of limited area may prove significant. It is important therefore that we test the validity of our assumption by calculating the effect of facet size on the extent of dipole coupling. This calculation is also relevant to the growth of islands of adsorbate on an extended surface and to adsorption on other small metal particles, such as supported catalysts.

Our model system consists of a small hexagonal island of the Cu(111) surface carrying a $\sqrt{3} \times \sqrt{3}$-30° overlayer of CO. The diameter of the island is increased progressively by adding successive hexagonal rings of thickness $2.56\sqrt{3}$ Å (i.e. the CO nearest neighbour distance) as shown in Fig.8.

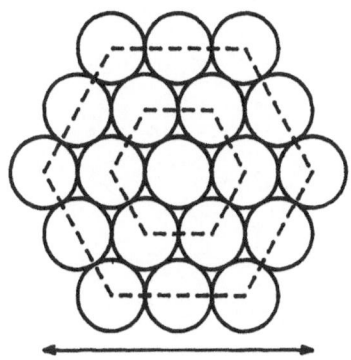

D

Fig.8 Hexagonal 19-molecule island

We assume that both the chemical shift and the distance to the image plane are independent of the island size. These assumptions appear reasonable for adsorption on an extended plane, but it is recognised that the image interaction is modified on very small particles [70].

In the case of the smaller islands, the solution of the full matrix eigenvalue problem is practicable, both for islands consisting of a single isotope and for those containing a mixture of ^{12}CO and ^{13}CO molecules [71]. As an example, Fig.9 shows spectra calculated for an assembly of 19-molecule islands (as in Fig.8) over the full range of composition from pure ^{12}CO to pure ^{13}CO (in 10% increments). It is seen that even in this very small island the typical effects of dipole coupling are evident. Two distinct bands are observed for each isotopic mixture, with a pronounced intensity transfer to the high frequency band. There is also a substantial shift in the position of the bands as the composition is varied; the extent of this shift is approximately 70% of that observed in an infinite overlayer.

Another effect seen in the spectra of small islands is a distinct shoulder on the low frequency side of the bands, most obvious in the single-isotope spectra. Because of the non-equivalence of the different molecules in the island certain vibrational modes other than the fully in-phase one become infrared active and give rise to the small low-frequency bands which produce these shoulders.

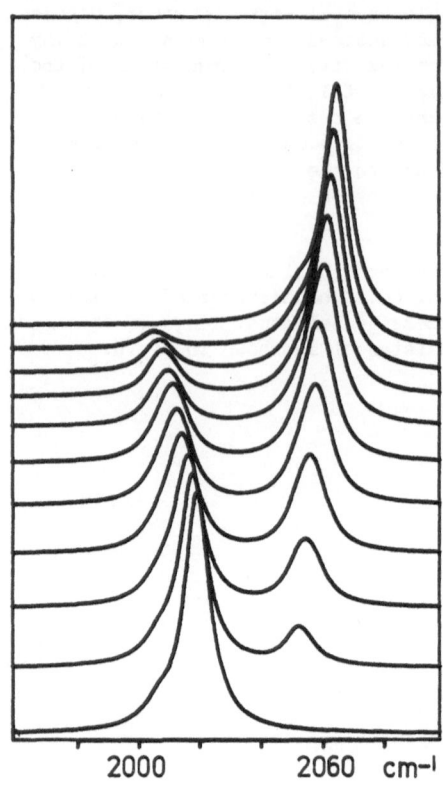

Fig.9 Computed spectra of assembly of 19-molecule islands as function of isotopic composition (0% ^{12}CO at bottom, 100% ^{12}CO at top)

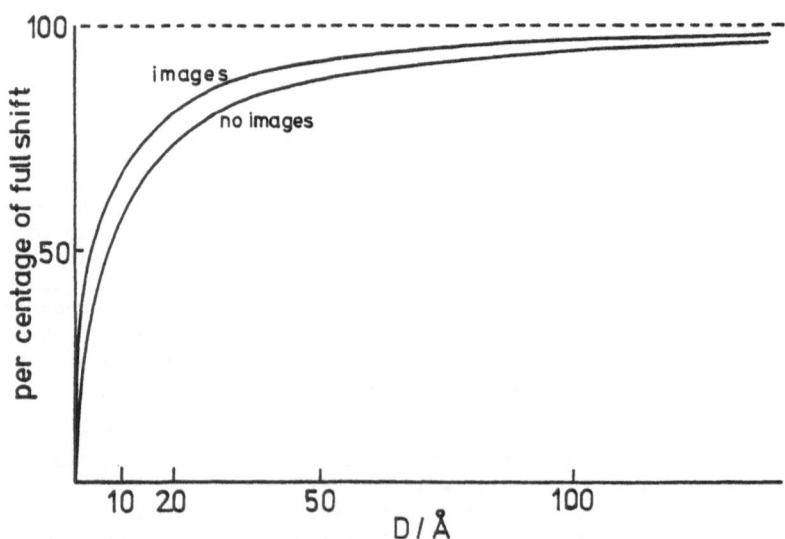

Fig.10 Dipole coupling shift relative to infinite layer for islands of size D

In the case of the larger islands, exact solutions to the eigenvalue problem were not sought, but instead the shift corresponding to the dipole sum at the central molecule was evaluated. Shifts calculated in this way are slightly larger than those which would be observed for the hexagonal island used in the model, but they should provide a reasonable estimate of the shift to be expected for an irregular island in which the typical distance of a molecule from the edge of the island is of the order of D/2. These calculated shifts are shown as a function of D in Fig.10. In order to demonstrate that the predicted influence of particle size is relatively insensitive to the details of the particular dipole-coupling model employed, we have calculated the shift using two different models - one which takes account of image interactions and uses the values of the molecular parameters found to be appropriate for CO/Cu(111), and one which neglects image effects entirely.

In each case it is seen that, although the dipole sum is indeed very slow to converge fully, the initial convergence is rapid and only the smallest islands will show marked differences from the infinite layer. The particles comprising the film whose spectrum is shown in Fig.4 are typically 100 nm in diameter, and it is clear that for particles of this size comparison with the spectra from extended single crystal surfaces is fully justified. In fact, even when the particle size is reduced to 3 nm the frequency difference amounts to only about half of the line width and is significantly less than the difference between the frequencies observed on different single crystal faces at comparable coverages. Supported copper catalysts tend to exhibit particle sizes rather greater than 3 nm, and so it seems that incomplete dipole sums are not a major factor in influencing IR band positions on this metal. However, the effect should be borne in mind when comparing spectra on other metals for which particle sizes may well be smaller.

Acknowledgements

We gratefully acknowledge the support of this work by the Science Research Council and the Central Research Fund Committee of London University. We thank Dr. B. N. J. Persson and Dr. K. P. de Jong for preprints of their papers.

1   R. P. Eischens, W. A. Pliskin, S. A. Francis: J. Chem. Phys. 22, 1786 (1954)
2   R. P. Eischens, S. A. Francis, W. A. Pliskin: J. Phys. Chem. 60, 194 (1956)
3   R. P. Eischens, W. A. Pliskin: Adv. Catalysis 10, 1 (1958)
4   N. Sheppard, T. T. Nguyen: In Advances in Infrared and Raman Spectroscopy, ed. by R. J. H. Clark and R. E. Hester (Heyden, London, 1978) Vol. 5
5   R. M. Hammaker, S. A. Francis, R. P. Eischens: Spectrochim. Acta 21, 1295 (1965)
6   R. F. Baddour, M. Modell, R. L. Goldsmith: J. Phys. Chem. 74, 1787 (1970)
7   A. Palazov, C. C. Chang, R. J. Kokes: J. Catalysis 36, 338 (1975)
8   T. Engel, G. Ertl: Adv. Catalysis 28, 1 (1979)
9   R. P. Eischens, J. Jacknow: In Proc. 3rd. Intern. Congr. Catalysis (North Holland, Amsterdam, 1965) p.627
10 R. van Hardeveld, A. van Montfoort: Surface Sci. 4, 396 (1966)
11 S. A. Francis, A. H. Ellison: J. Opt. Soc. Amer. 49, 131 (1959)
12 R. G. Greenler: J. Chem. Phys. 44, 310 (1966)
13 J. D. E. McIntyre: In Advances in Electrochemistry and Electrochemical Engineering, ed. by R. H. Muller (Wiley, New York, 1973) Vol. 9, p.61
14 H. G. Tompkins: In Methods of Surface Analysis, ed. by A. W. Czanderna (Elsevier, New York, 1975) p.447

15 J. Pritchard, T. Catterick: In Experimental Methods in Catalytic Research, ed. by R. B. Anderson and P. T. Dawson (Academic Press, New York, 1976) Vol. 3, p.281
16 J. Pritchard: In Chemical Physics of Solids and Their Surfaces, ed. by M. W. Roberts and J. M. Thomas (Specialist Periodical Reports, The Chemical Society, London, 1978) Vol. 7, p.157
17 M. J. Dignam, J. Fedyk: Appl. Spectroscopy Rev. 14, 249 (1978)
18 G. W. Poling: J. Colloid Interface Sci. 34, 365 (1970)
19 K. Horn, J. Pritchard: Surface Sci. 52, 437 (1975)
20 P. Hollins, J. Pritchard: Surface Sci. 89, 486 (1979)
21 J. F. Blanke, S. E. Vincent, J. Overend: Spectrochim. Acta 32A, 163 (1976)
22 R. G. Greenler: J. Vac. Sci. Technol. 12, 1410 (1975)
23 F. M. Hoffmann, A. M. Bradshaw: In Proc. 7th Intern. Vac. Congr. and 3rd Intern. Conf. Solid Surfaces, ed. by R. Dobrozensky, F. Ruedenauer and F. P. Viehboeck (R. Dobrozensky, Vienna, 1977) Vol. 2, p.1167
24 P. E. Wierenga, G. J. Mollenhorst, A. T. B. Witterink: Rev. Sci. Instrum. 49, 408 (1978)
25 M. L. Kottke, R. G. Greenler: Rev. Sci. Instrum. 42, 1235 (1971)
26 R. W. Roberts, J. F. Harrod, H. A. Poran: Rev. Sci. Instrum. 38, 1105 (1967)
27 P. Hollins, J. Pritchard: J. Vac. Sci. Technol. 17, 665 (1980)
28 H. G. Tompkins, D. L. Allara: Rev. Sci. Instrum. 45, 1221 (1974)
29 H. J. Krebs, H. Lüth: Appl. Physics 14, 337 (1977)
30 M. Moskovits, C. J. Hope, B. Jantzi: Canad. J. Chem. 53, 3313 (1975)
31 A. M. Bradshaw, F. M. Hoffmann: Surface Sci. 72, 513 (1978)
32 M. J. Dignam, B. Rao, M. Moskovits, R. W. Stobie: Canad. J. Chem. 49, 1115 (1971)
33 R. W. Stobie, B. Rao, M. J. Dignam: Surface Sci. 56, 334 (1976)
34 R. W. Stobie, B. Rao, M. J. Dignam: J. Opt. Soc. Amer. 65, 25 (1975)
35 J. D. Fedyk, P. Mahaffy, M. J. Dignam: Surface Sci. 89, 404 (1979)
36 W. G. Golden, D. S. Dunn, J. Overend: J. Phys. Chem. 82, 843 (1978)
37 W. G. Golden, D. S. Dunn, C. E. Pavlik, J. Overend: J. Chem. Phys. 70, 4426 (1979)
38 D. S. Dunn, W. G. Golden, M. W. Severson, J. Overend: J. Phys. Chem. 84, 336 (1980)
39 J. Pritchard, M. L. Sims: Trans. Faraday Soc. 66, 427 (1970)
40 H. G. Tompkins, R. G. Greenler: Surface Sci. 28, 194 (1971)
41 M. Grunze, R. K. Driscoll, G. N. Burland, J. C. L. Cornish, J. Pritchard: Surface Sci. 89, 381 (1979)
42 R. J. Behm , K. Christmann, G. Ertl, M. A. van Hove, P. A. Thiel, W. H. Weinberg: Surface Sci. 88, L59 (1979)
43 F. M. Hoffmann, A. M. Bradshaw: J. Catalysis 44, 328 (1976)
44 R. A. Shigeishi, D. A. King: Surface Sci. 58, 379 (1976)
45 A. Crossley, D. A. King, Surface Sci. 68, 528 (1977)
46 P. Hollins, J. Pritchard: In Vibrational Spectroscopies Applied to the Characterization of Adsorbed Species on Catalysts (ACS Symposium Series, American Chemical Society, Washington, 1980). In press
47 K. P. de Jong, J. W. Geus, J. Joziasse: Applic. Surface Sci. In press
48 J. Pritchard, T. Catterick, R. K. Gupta: Surface Sci. 53, 1 (1975)
49 M. McLean, B. Gale: Phil. Mag. 20, 1033 (1969)
50 J. C. Decius, R. M. Hexter: Molecular Vibrations in Crystals (McGraw Hill, New York, 1977)
51 R. M. Hexter: J. Chem. Phys. 33, 1833 (1960)
52 D. A. Dows: J. Chem. Phys. 32, 1342 (1960)
53 A. Crossley, D. A. King: Surface Sci. 95, 131 (1980)
54 K. Horn, J. Pritchard: J. Phys. (Paris) 38, C4-164 (1977)

55 B. N. J.Persson, R. Ryberg: to be published
56 R. van Hardeveld, A. van Montfoort: Surface Sci. 17, 90 (1969)
57 G. N. Burland, J. Pritchard: unpublished results
58 T. L. Brown, D. J. Darensbourg: Inorg. Chem. 6, 971 (1967)
59 G. D. Mahan, A. A. Lucas: J. Chem. Phys. 68, 1344 (1978)
60 M. Scheffler: Surface Sci. 81, 562 (1979)
61 A. M. Bradshaw, M. Scheffler: J. Vac. Sci. Technol. 16, 447 (1979)
62 H. Pfnür, D. Menzel, F. M. Hoffmann, A. Ortega, A. M. Bradshaw: Surface
    Sci. 93, 431 (1980)
63 P. Hollins, J. Pritchard: Chem. Phys. Lett., in press
64 P. Hollins: Ph.D. Thesis (University of London, 1980)
65 S. Andersson J. B. Pendry: Phys. Rev. Lett. 43, 365 (1979)
66 P. J. Jennings, G. L. Price: Surface Sci. 93, L124 (1980)
67 P. Hollins, J. Pritchard: Surface Sci., in press
68 M. Moskovits, J. E. Hulse: Surface Sci. 78, 397 (1978)
69 S. Efrima, H. Metiu: Surface Sci. 92, 433 (1980)
70 M. Cini: Surface Sci. 62, 148 (1977)
71 P. Hollins: unpublished results

9. Raman Spectroscopy of Adsorbates at Metal Surfaces

J. A. Creighton

With 12 Figures

9.1 Background

Many of the important problems of applied significance in surface chemistry involve surfaces in contact with high pressure gases or with liquids or solids. These interfaces, which include catalyst surfaces under manufacturing conditions and electrochemical interfaces, cannot be studied directly by the high vacuum techniques of electron energy loss (EELS), Auger, or photoelectron spectroscopy, or by secondary ion mass spectrometry, and much interest has therefore recently been shown in developing Raman spectroscopy for surface molecular studies. This technique, which involves inelastic scattering of visible photons and is well known as a technique for vibrational spectroscopy in other fields, has several experimental features which make it particularly attractive for surface studies: (1) there are no limitations on the phase in contact with the surface other than that it is transparent to the incident and scattered photons. The ambient phase may therefore be a condensed phase or a gas, or vacuum; (2) the laser light source and spectrometer are external to the apparatus, the incident and scattered light entering and leaving the sample chamber through glass windows. The technique thus makes minimal demands on the design of the sample chamber; (3) the focussed laser beam provides microprobe capability; (4) the frequency resolution is good, being typically in the range 1-10 cm^{-1} depending on the slit width necessary for adequate signal/noise ratio. It is thus comparable in resolution to surface infrared spectroscopy and much better than EELS and photoelectron or Auger spectroscopies.

It has however been generally accepted that Raman spectroscopy is a technique of only moderate sensitivity. This is true if the incident photons are of frequency well away from the electronic absorption frequencies of the scattering species, when it is somewhat less sensitive than infrared spectroscopy and therefore several orders of magnitude less sensitive than EELS. Thus the cross sections for non-resonant Raman scattering are typically 10^{-30} cm^2 molecule^{-1} steradian^{-1} [1]. A simple calculation shows that for a surface coverage of 10^{14} molecules cm^{-2} and a laser power of 100mW, and taking account of the spectrometer collection angle of 0.1 steradians and the 20% detector sensitivity, the Raman band photon count rate is $ca.$ 0.5 Hz. The background count rate in surface Raman studies is typically $ca.$ 10^3 Hz, and the detection of monolayers of molecules on surfaces by Raman spectroscopy away from resonance is thus not a good prospect.

This situation may be transformed if use is made of resonant Raman scattering. It has long been known that if the wavelength of the incident light is chosen so as to lie in a region of electronic absorption by the scattering molecules, there may be an enhancement of the Raman scattering intensity by several orders of magnitude. Interest in exploiting this enhancement has begun to develop in a number of fields, particularly for the investigation of substances of biological importance where ultraviolet as well as visible irradiation has been used [2]. For obvious practical reasons however attention has been mainly directed towards substances with absorptions in the visible region, and enhancements of up to 10^6-fold have been observed for some dyes [3]. Examples where this enhancement has been exploited to enable Raman spectra of adsorbed dye molecules to be obtained are the investigations of rose Bengal or methylene blue at metal oxide electrodes [4], or of methyl orange at a carbon tetrachloride-aqueous surfactant interface [5].

The purpose of this review is to outline some features of a remarkable new resonance Raman phenomenon which has been found to occur for molecules at some metal surfaces. This phenomenon, which has been referred to in the literature as surface-enhanced Raman scattering (SERS) and has recently been reviewed also by VAN DUYNE [6], has two features which are of particular significance in its application as a surface spectroscopic technique. Firstly although the laser wavelengths which have been used have so far been in the visible region, the method is in no way restricted to coloured substances. Indeed as will be shown, the range of adsorbates which give excellent quality SERS spectra is now quite varied, and there appears to be no limitation on this range other than that the adsorbate adsorbs under the prevailing ambient conditions. Secondly it appears that the resonant response is in the interface itself, and thus only the scattering by the adsorbed monolayer is enhanced. This is of great importance in the application of this technique to metal-condensed phase or metal-high pressure gas interfaces, since without it the Raman spectrum would be overwhelmingly dominated by bands of the more abundant ambient phase. This review is not exhaustive, but is intended rather to give an indication of the quality of surface vibrational spectra which may be obtained by this method, and to present some results which bear on the SERS selection rule and on the mechanism of this phenomenon. It will be apparent that this new technique is still at the stage of early exploration. The relatively complex organic adsorbates which have mainly been studied have been chosen for their adsorption affinities and not for their chemical interest, and detailed vibrational analyses have not been made. Its application to simple adsorbate-metal systems investigated by other surface vibrational techniques and the study of electrochemical and other surface reactions thus remains to be developed, but already the technique shows great promise as a tool in surface chemistry, and it has attracted considerable recent interest [7].

9.2 Experimental Features

Raman spectra of metal-adsorbed monolayers showing good signal/noise ratios were first reported by FLEISCHMANN, HENDRA and McQUILLAN, who investigated pyridine adsorbed on a silver electrode in an aqueous electrochemical cell [8]. The silver electrode had been electrochemically roughened by current-cycling in aqueous chloride ion solution in order to increase its surface area. In repeating this work it was recognized by JEANMAIRE and VAN DUYNE [9] and by ALBRECHT and CREIGHTON [10] that optimization of this surface roughening gave very large adsorbate Raman band intensities which were much greater than could be accounted for by the increase in surface area. Recent measurements of the amount of adsorbed pyridine or other adsorbates indicate that the surface roughness factor for these silver surfaces is between one and several hundred, depending on the amount of charge passed in the roughening cycle [11] and it

is clear therefore that there is a surface Raman enhancement due to some
other cause of about 10^4, giving a total enhancement relative to a smooth
surface monolayer in the region of 10^4-10^6.

Most, though not all, of the published work on SERS has been on adsorbates
at aqueous silver electrodes roughened in this way. The experimental arrangement is very straightforward, and Fig. 1 shows a diagram of the electrochemical
cell originally used by HENDRA and coworkers [8]. A polycrystalline silver
rod is illuminated on its end face at an angle of incidence of about 60° by
a focussed laser beam, which enters through the flat end window of the glass
cell. The area of the surface illuminated by the focussed laser is imaged
through the end window onto the spectrometer entrance slit with the usual
large aperture collection lens arrangement so as to collect the Raman-
scattered light, which is emitted in a relatively narrow cone [12,13]. The
cell is filled with electrolyte containing the adsorbate, and the silver rod
is sleeved in teflon so that only the end face is exposed to electrochemical
action. The cell carries secondary and reference electrodes for potentio-
static or other electrochemical control of the metal electrolyte interface.

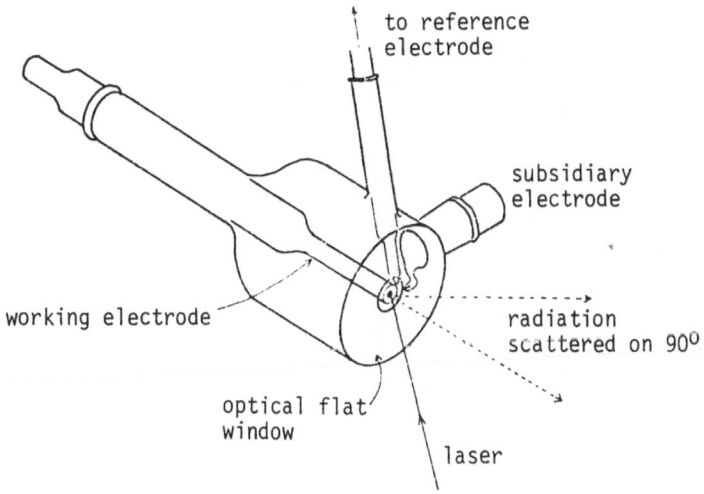

Fig. 1. An electrochemical cell for surface Raman spectroscopy (from [8]).

The metal surface is usually cleaned by abrading with fine emery, which
leaves it relatively coarsely furrowed on an optical scale, and it may then
be polished. If this surface is placed in a dilute aqueous pyridine solution
(10^{-2}M pyridine, 10^{-1}M KCl) at open circuit potential and illuminated by the
laser, the Raman spectrum shows weak bands characteristic of the dilute
solution pyridine excited by the light as it passes through the electrolyte
on its way to the silver surface. At very high gain bands due to adsorbed
pyridine are also discernible but they are very weak, although it is known
that there is saturated monolayer coverage of silver at this pyridine solution concentration. It is only when the surface is roughened that intense
Raman bands of adsorbed pyridine with photon count rates of up to 10^5 Hz are
observed. This surface roughening is most effectively done by forming *ca.* 100
atomic layers of AgCl on the surface, by for example taking its potential to
+ 150 mV (SCE) for 20 sec during which time charge equivalent to *ca.* 50 mC

cm^{-2} is passed, and then stepping to -50 mV, whereupon the granular AgCl film is reduced back to metallic silver with almost complete charge recovery. This procedure, which takes $ca.$ 30 sec, has a most dramatic effect on the Raman spectrum. As Fig. 2 shows, the intense Raman spectrum of adsorbed pyridine begins to appear at the onset of the reduction current as reformed silver begins to form, reaching a maximum when the reduction is substantially complete [10]. Part of the resulting surface pyridine Raman spectrum is shown in Figs. 3 and 8. The possibility that the intense spectrum is due to a relatively thick layer of a silver pyridine complex deposited on the surface may be discounted from the fact that there is virtually complete reduction (to within 99.9%) of the $ca.$ 100 atomic layers of AgCl formed in the roughening cycle [6] and that similar high intensities are obtained if pyridine is added after rather than before the roughening cycle. The reformed silver surface has been thoroughly examined by electron microscopy, Auger spectroscopy and SIMS and shown to be exceedingly rough and porous and chemically very pure [14]. Other electrolyte anions including F$^-$, Br$^-$, I$^-$, [SO$_4$]$^{2-}$ and [ClO$_4$]$^-$ have also been successfully used in place of chloride in the roughening cycle.

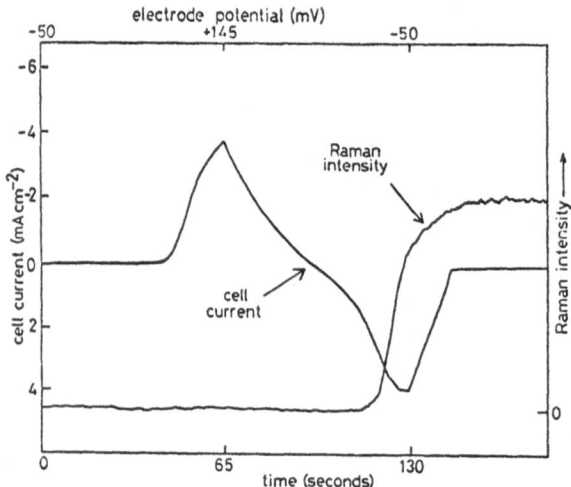

Fig. 2. The cell current and intensity of the 1025 cm^{-1} Raman band during a roughening cycle involving a linear voltage sweep at a silver electrode in 0.1M KCl, 0.01M pyridine.

9.3 Enhanced Raman Spectra of Adsorbates on Silver Surfaces

The Raman spectrum of pyridine adsorbed on silver has been discussed briefly by several authors and reviewed by VAN DUYNE [6], but it is clear that the identity of the surface species are far from settled. In the presence of aqueous chloride electrolyte the spectrum is dependent on the electrode potential, as originally noted by HENDRA and coworkers [8]. Figure 3, which is reproduced from [8], shows the variation with potential of the relative intensities of bands in the region of 1000 cm^{-1}. It is clear from variations here and throughout the spectrum that there are two quite distinct surface pyridine species. There are also small shifts of <5 cm^{-1} in the frequencies

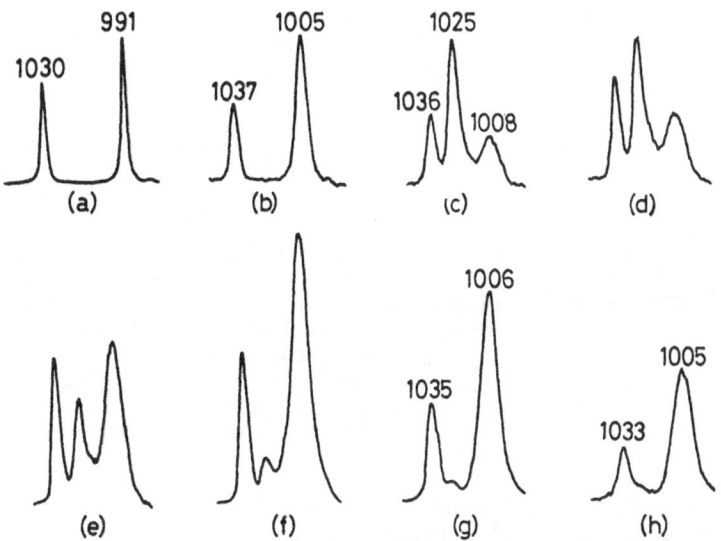

Fig. 3 Raman spectra of pyridine in solution and at a silver electrode in 0.1M KCl; (a) liquid pyridine; (b) 0.05M aqueous pyridine; (c) silver electrode 0.0 V (SCE); (d) -0.2 V; (e) - 0.4 V; (f) -0.6 V; (g) -0.8 V; (h) -1.0 V. Frequencies given in cm^{-1}. From [8].

of some of the bands with change of potential [15], small frequency differences for pyridine adsorbed at different silver single crystal faces which have been minimally roughened [16], and small random variations in the frequency of the relatively broad band near 1008 cm^{-1} for different polycrystalline electrodes, which may indicate several different adsorption sites for these surface pyridine species. Though much further work is required to identify these species firmly, it appears at present that one of them consists of pyridine molecules directly bound through nitrogen to the silver surface. This adsorbed pyridine, which is the only form detected on silver by SERS if the electrode potential is in the range -0.6 to -1.0 V or if chloride or bromide ions are absent from the electrolyte, gives rise to prominent SERS bands at 1035 and 1006 cm^{-1} (Fig. 3) together with other weaker bands (Fig. 8). It is of interest that these frequencies are very close to the frequencies of the Ag(I) complex [Agpy$_2$]$^+$. The possibility that this complex is the adsorbed species however is ruled out by the fact that the 1035 and 1006 cm^{-1} SERS bands reach maximum intensity at about -0.6 V and persist to -1.1 V [8], at which electrode potentials reduction of this complex to Ag(0) would have occurred. Particularly interesting in the SERS spectrum for this surface species is a prominent band at 216 cm^{-1}, well below any internal vibration frequencies of pyridine and almost certainly due to ν(Ag-N) [6], and even more remarkable is a very intense SERS band at 8 cm^{-1} attributed to adsorbed pyridine libration [17]. The a_1 and b_2 ν(Ag-N) frequencies of [Ag py$_2$]$^+$ (assumed to be D_{2d}) are 250 and 245 cm^{-1} respectively.

The other surface pyridine species, which SERS shows to be abundant on silver at potentials between 0 and -0.4 V *and* in the presence of chloride or bromide ions, is characterised by a strong band at 1025 cm^{-1} (Fig. 3), and by other bands. The identity of this species is less certain. However it is notable that its bands are particularly intense if the electrolyte is saturated with AgCl, thus forming Ag(I) chloro-complexes in solution, and we are therefore of the opinion that it is an adsorbed complex $[Ag\ py_mCl_n]^{(n-1)-}$ (m, n = 1 or 2). The 1025 cm^{-1} SERS band is very close to the frequency of the most intense pyridine ring mode of the known pyridine chloro-complexes $[Znpy_2Cl_2]$ and $[Rhpy_4Cl_2]^+$, and reduction or desorption of this silver chloro-complex (which has a negative charge if n = 2) would account for the loss of the bands of this species at negative electrode potentials. Our recent observation of the SERS spectrum of the non-labile complex $[Rhpy_4Cl_2]^+$ adsorbed on silver further supports this proposal. The mode of adsorption of this and other chloro-complexes is presumed to be through a bridging function of the chloride ligand. Such halide bridges between two metal atoms are well known in inorganic chemistry where they play an important role in electron transfer and ligand exchange, and their participation in binding metal ion complexes to metal surfaces may account for the well documented phenomenon of chloride-induced ligand adsorption. The 240 cm^{-1} band which is a prominent feature of SERS spectra of silver surfaces roughened in chloride electrolyte (in the presence or absence of other adsorbates) may be due to the stretching of such a bridge between the surface and a silver ion.

Roughened silver surfaces bearing other adsorbates may be prepared as they are for pyridine by electrochemical roughening in the presence of the adsorbate, or by displacement of adsorbed pyridine. Aqueous or non-aqueous solutions of adsorbates give equally satisfactory results, though there may be difficulties from impurities in organic solvents. A wide range of adsorbates have now been reported to give SERS spectra at silver surfaces, including several unsaturated and saturated nitrogen bases [9, 15, 18]; cyanide ion [19, 20], thiocyanate ion [21], tetra-alkylammonium ions [22], triphenylphosphine and triphenylarsine, ethanethiolate, benzoate, $[Rhpy_4Cl_2]^+$ and $[Ru_3O_2(NH_3)_{14}]^{6+}$ ion [23], p-nitroso-dimethylaniline [24], and methyl orange and crystal violet [9]. The following spectra are presented to illustrate the quality of the data

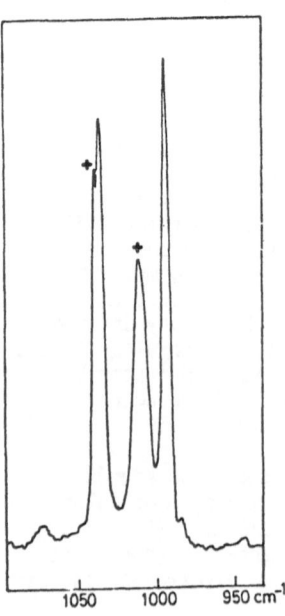

Fig. 4. Raman spectrum of roughened silver electrode immersed in liquid pyridine. Bands marked + are due to adsorbed pyridine, other two bands are due to liquid pyridine.

Figure 4 is a reproduction of a part of the spectrum of a roughened silver electrode immersed in liquid pyridine. The Raman bands of liquid pyridine at 1030 and 991 cm^{-1} are well known to be exceedingly intense, yet the bands of the surface pyridine monolayer are of comparable intensity. This is a striking illustration not only of the magnitude of the surface Raman enhancement but also of its selectivity towards only the adsorbed monolayer. It is clear that were the enhancement to extend only a few monolayers out from the surface, the bands of the bulk liquid pyridine would dominate the spectrum.

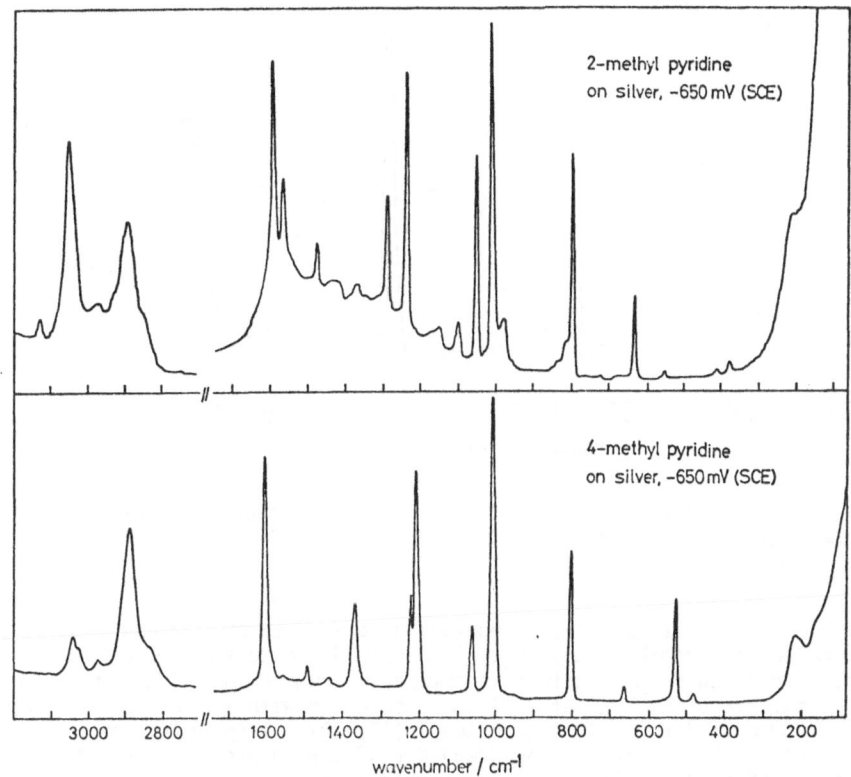

Fig. 5. Raman spectra of 2-methyl and 4-methylpyridines at the silver/0.1M aqueous KCl interface.

Figure 5 shows the SERS spectra of two of the methyl pyridines. The frequencies are close to those of the free molecules in aqueous solution but there are substantial relative intensity differences, notably in the bands near 800 cm^{-1} due principally to ring-CH$_3$ stretching, which is very weak in the SERS spectrum of the 3-methyl compound. Exposure of silver surfaces bearing these relatively labile adsorbates to a solution of ethanethiol, or roughening a silver surface in the presence of this substance, yields the spectrum in Fig. 6. This surface Raman spectrum resembles the spectrum of liquid ethanethiol except for small shifts in the frequencies and relative intensities, the intense ν(C-S) band at 639 cm^{-1} in the SERS spectrum lying at 658 cm^{-1}

in liquid ethanethiol. It is notable however that the ν(S-H) band, which is a conspicuous band at 2571 cm⁻¹ in the Raman spectrum of liquid ethanethiol, is absent from the SERS spectrum, and it is therefore clear that the adsorbed species is the ethanethiolate ion $C_2H_5S^-$, doubtless bound to the surface through the sulphur atom as in the known thiolate complexes of Ag(I).

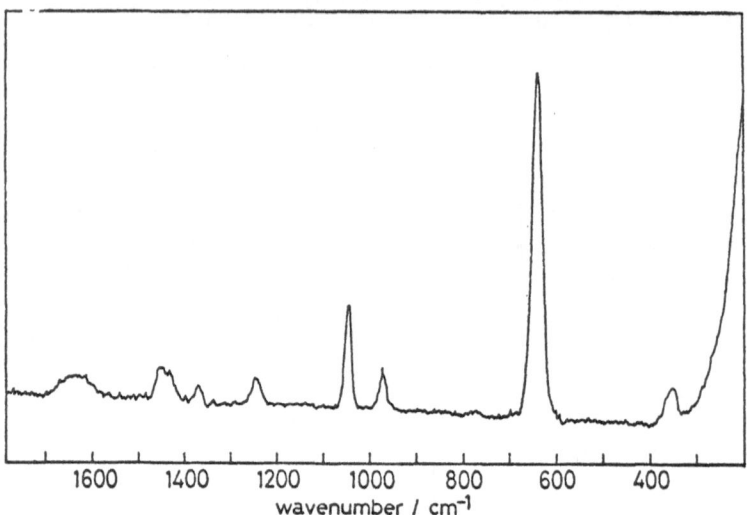

Fig. 6. Raman spectrum of the ethanethiolate ion adsorbed at a silver/ saturated aqueous ethanethiol interface at open circuit potential.

There is at present very little published Raman data on significantly smaller molecules adsorbed on silver. The only diatomic species which has provided data at an aqueous silver interface is cyanide ion. This shows an extremely intense ν(CN) band at 2114 cm⁻¹, together with a moderately strong band near 230 cm⁻¹ due possibly to ν(Ag-C) [19, 20]. This ν(CN) frequency is very close to the corresponding frequency of the complex ion $[Ag(CN)_3]^{2-}$, though the SERS band envelope is sufficiently broad that it almost encompasses also the frequencies of $[Ag(CN)_2]^-$ and $[CN]^-$. OTTO [20] has therefore argued that the principal surface species is an $[Ag(CN)_3]^{2-}$ surface complex bound to the surface through the silver atom. The oxidation state of the silver atom is thus not defined, and the surface complex may be regarded as a silver adatom bearing several adsorbate ions. It is of note that for other adsorbates, including pyridine, the surface Raman frequencies are also very close to those of known Ag(I) complexes, and OTTO has proposed that such silver adatoms in surface complexes may be of crucial significance to SERS in providing surface roughness on an atomic scale [20].

Evidence has been sought for enhanced scattering by water molecules adsorbed at aqueous silver interfaces. There is the difficulty here that the Raman bands from the aqueous phase are relatively broad and would overlie the SERS bands of adsorbed water. By analogy with the observations for silver in liquid pyridine however (Fig. 4), the SERS bands would be of comparable intensity to the Raman bands of bulk water. The presence of a significant contribution to the Raman spectrum of an aqueous silver surface from SERS bands of adsorbed water might therefore be apparent through an anomalously high value of the depolarization ratio of the polarized $v_1/2v_2$ water band at $ca.$ 3230 cm^{-1}, since it appears to be generally the case that SERS bands are approximately depolarized. In our own work on silver electrodes roughened by cycling in 0.1M KCl and examined in this electrolyte however, the depolarization ratios of the water bands were found to be normal.

Surface Raman studies of such relatively weakly adsorbing molecules may best be carried out on cooled specimens under UHV conditions, and there can be little doubt that investigations of this kind will be developed in several laboratories in the very near future. Already WOOD and KLEIN have published results showing several intense surface Raman bands in the $v(CO)$ region from submonolayer CO adsorbed on cooled polycrystalline silver films [25]. The multiplicity of bands in this region and the differences in the pattern of $v(CO)$ bands for different surfaces or with changes in CO coverage clearly indicate multiple adsorption sites, with CO bound both terminally and in a bridging configuration. Several bands of lower frequency were also observed and were attributed mainly to impurities. The silver films used in this work were prepared by evaporation onto abrasively polished copper substrates and were therefore considerably less rough than the aqueous silver surfaces prepared electrochemically. Nevertheless the Raman intensity enhancement due to the metal surface was estimated to be 10^5-fold from measurements of the amount of adsorbed CO by flash desorption. A UHV Raman study of pyridine adsorbed on silver near 70K has also been reported [26]. In this study a bulk polycrystalline silver sample was abrasively polished and was then subjected to argon ion beam sputtering before exposure to pyridine. These surfaces were thus also much less rough than those produced electrochemically, and the SERS spectra were apparently $ca.$ 1000-fold weaker than those which may be obtained from pyridine at the electrochemically roughened silver surface. A surprising result from this study however is the observation that the adsorbed pyridine frequencies are almost identical to those of the pure substance, and the pyridine thus appears to be physisorbed under these conditions. This is in contrast to the results for the aqueous silver interface, noted above, where there were substantial shifts in the pyridine frequencies accompanying adsorption.

9.4 Enhanced Raman Spectra of Adsorbates at Metals Other Than Silver

In view of these promising results for adsorbates on silver surfaces it is of obvious importance to evaluate other metals for surface-enhanced Raman scattering. Success here has so far been limited to very few other metals however, at least using similar roughening and excitation conditions to those used for silver surfaces. Following an early report of moderately intense Raman scattering by pyridine adsorbed on an aqueous copper electrode by HENDRA and coworkers [27], WENNING, PETTINGER and WETZEL [13] have recently given Raman data for pyridine adsorbed on aqueous copper or gold surfaces which had been roughened by an electrochemical oxidation-reduction cycle in aqueous KCl. It was found to be very important with these metals to use red excitation, and whereas with 568.2 nm excitation little or no surface intensity enhancement was observed, with 647.1 nm excitation the surface pyridine spectra were of good intensity with estimated enhancement factors of 10^5 and 10^4 for copper and gold respectively [13]. With slightly different surface roughening procedures we have observed at least a 10^6-fold total intensity enhancement for pyridine on copper with 647.1 nm excitation, giving a Raman photon count rate of over 10^5 Hz for the strongest band at 1015 cm^{-1} (Fig. 7) for 100 mW incident laser power. The adsorbed pyridine spectrum shows no evidence of two surface pyridine

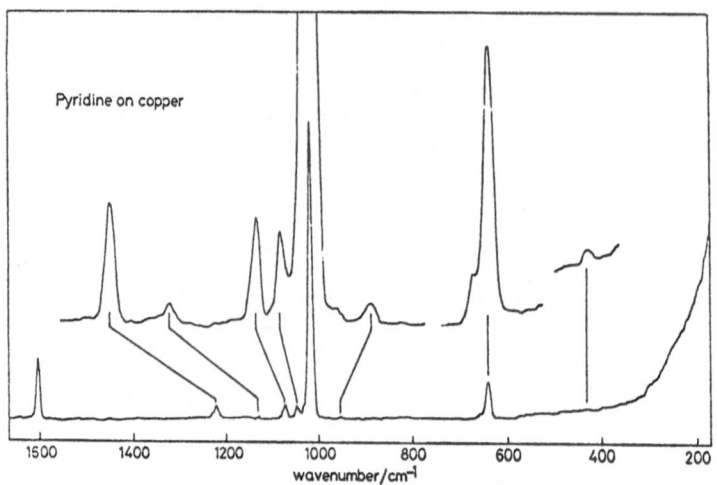

Fig. 7. Raman spectrum of pyridine adsorbed at a copper electrode in 0.1M aqueous NaCl, -500 mV (SCE).

species as was observed for silver surfaces in aqueous electrolytes containing chloride ions, and by comparing the frequencies of pyridine on copper with those for silver surfaces it appears that the pyridine is directly bound to the copper surfaces through nitrogen coordination. It is of interest that this has been shown to be the mode of attachment of pyridine to copper (110) by angle-resolved photoelectron emission [28]. Intense SERS spectra of triphenylphosphine at aqueous copper surfaces, and of pyridine [29] and ethanethiolate ions on aqueous colloidal gold particles have also been obtained with red excitation. These very recent results leave little doubt that, as for silver, much will be learned about the surface chemistry of copper and gold by Raman spectroscopy in due course.

Surface Raman studies of molecules at the surfaces of metals outside the Cu, Ag, Au subgroup have so far been less successful. Measurable bands have been reported for I_2 and CO at aqueous platinum electrodes [30, 31] and for CO and hydrogen at moderate pressures on nickel surfaces [32, 33]. The photon count rates with 514.5 nm excitation were in the region of 10-100 Hz and signal averaging or multichannel detection were used to improve the weakest signals [31, 33]. Assuming the nickel single crystal surfaces used in one of the studies [32] to be smooth it appears that there may have been a surface Raman enhancement for nickel of about 10-fold, but for the platinum surfaces any enhancement other than that due to the large surface area of platinum grey must have been smaller than this. There is clearly a great need outside the Cu, Ag, Au subgroup therefore for developments in the theoretical understanding of the surface Raman enhancement, which may lead to improvements in the surface preparation and excitation procedures for these other metals.

9.5 The SERS Selection Rule

The spectral activities of vibrations of molecules adsorbed at metal surfaces may differ from those of molecules in bulk samples, and the surface selection rules which govern these activities are of obvious importance for spectral assignments and for inferring molecular orientations on surfaces. Thus in metal surface infrared spectroscopy only vibrations with transition dipole components perpendicular to the surface are spectrally active [34]. In the specular component of EELS there is a similar rule which allows the observation only of these perpendicular dipolar modes, whereas the nonspecular component of EELS is not thus restricted [35].

The general selection rule for the enhanced Raman scattering at metal surfaces is not yet known. This is due to the present lack of knowledge of the mechanism of coupling of the adsorbate vibrations to the incident photons, since the direction of the electromagnetic field at metal surfaces of any shape and incident wavelength is known, or is at least calculable, from classical electromagnetic theory. At this stage of development the position is therefore one of empirical observations on SERS activities for specific molecules. Pyridine is a particularly useful molecule for exploring such activities since in the free state it has Raman-active vibrations in all four symmetry classes of the C_{2v} point group. Figure 8 shows part of the Raman spectrum of pyridine in aqueous solution with the symmetry classes of the fundamentals [36] indicated, and for comparison a SERS spectrum of pyridine on silver in aqueous KCl at -500 mV, where the pyridine is believed to be bound to the surface thro the nitrogen atom and thus retains its C_{2v} symmetry. It is seen that all the fundamentals discernible in the solution spectrum are present in the SERS spectrum, including the non-dipole active a_2 vibrations. There is therefore no SERS selection rule for this molecule which is more restrictive than the normal Raman selection rule for free pyridine molecules. The bands observed in the SERS spectrum of pyrazine adsorbed on silver are also in accordance with the normal Raman selection rule, the Raman-forbidden u modes of the free molecule becoming active in the surface spectrum due to the loss of the centre of symmetry of the molecule upon adsorption [37].

At present therefore there is no experimental evidence for a special surface Raman selection rule applicable to SERS spectroscopy. Some interesting observations relating to band intensities have been made however by ALLEN and VAN DUYNE [38], who noted the striking decrease in the intensity of the $\nu(CN)$ band near 2240 cm^{-1} of the series of molecules 4-cyano, 3-cyano, and 2-cyanopyridine adsorbed on silver (Fig. 9). This decrease was not seen in the Raman spectra of the free cyanopyridines, where these bands all have similar intensities. Intensity differences are also seen in the bands of the adsorbed methylpyridines near 800 cm^{-1}, due predominantly to ring-CH$_3$ stretching [23], but here it is the 3-methyl isomer for which this band is particularly weak. These relative intensity differences may be a consequence of the different bond orientations with respect to the surface, as ALLEN and VAN DUYNE suggested [38]. It should be noted however that merely an orienting effect on the molecules by the surface cannot itself account for the $\nu(CN)$ intensity differences observed for the cyanopyridines, since the polarizability derivatives on which the Raman intensities depend are tensor quantities which do not necessarily vary much with orientation. This is certainly so for the $\nu(CN)$ modes of the cyanopyridines, at least in solution, since it is clear from the small Raman depolarization ratios of the $\nu(CN)$ bands of these molecules that the polarizability derivative tensors are near-isotropic.

An attempt to derive a general symmetry-based selection rule for SERS by considering the symmetry properties of the polarizability of an adsorbed molecule-image pair at a metal surface has been made by HEXTER and ALBRECHT [39]. It was concluded that any vibrations belonging to the same symmetry species as x^2, y^2, z^2 and xy are metal surface-Raman active, where z is perpendicular to the surface. For adsorbed N-bonded pyridine only the a_1 and a_2 vibrations are therefore predicted to be surface-Raman active. This is contrary to the experimental findings, as Fig. 8 shows. The image-field approach applied to absorption at infrared frequencies successfully yields

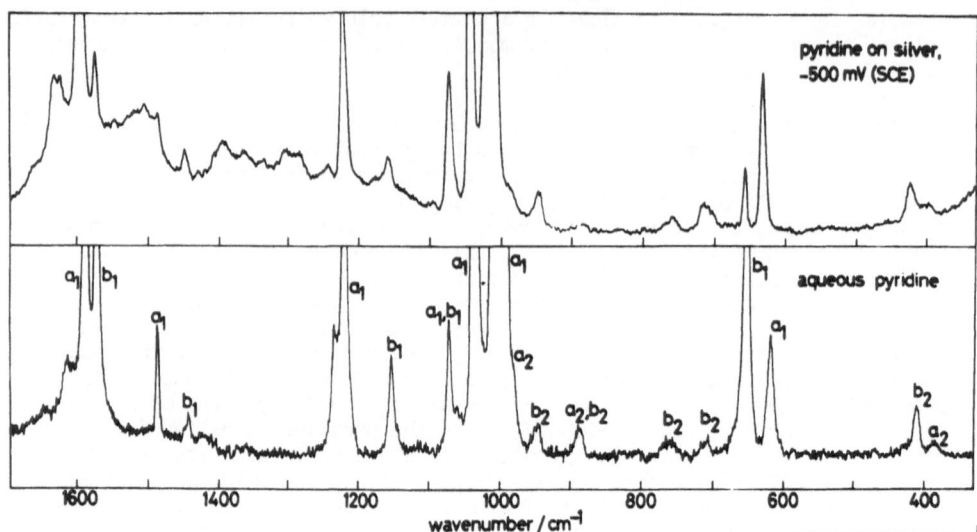

Fig. 8. The Raman spectra of pyridine adsorbed on silver in aqueous 0.1M KCl, and of a 2M aqueous pyridine solution, showing the assignment of fundamentals taken from [36]. Laser power 200 mW (514.5 nm), spectral band pass 2 cm⁻¹.

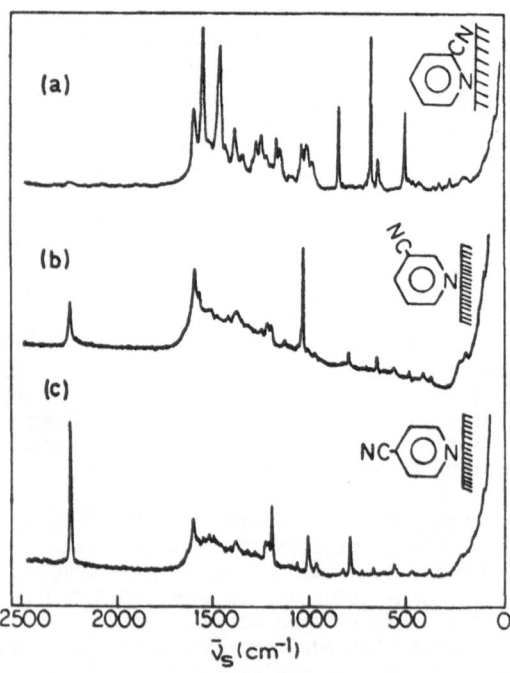

Fig. 9. Raman spectra of (a) 2-cyano-; (b) 3-cyano-; (c) 4-cyanopyridine at a silver electrode (0.6V SCE) in 0.1M aqueous KCl. From [38].

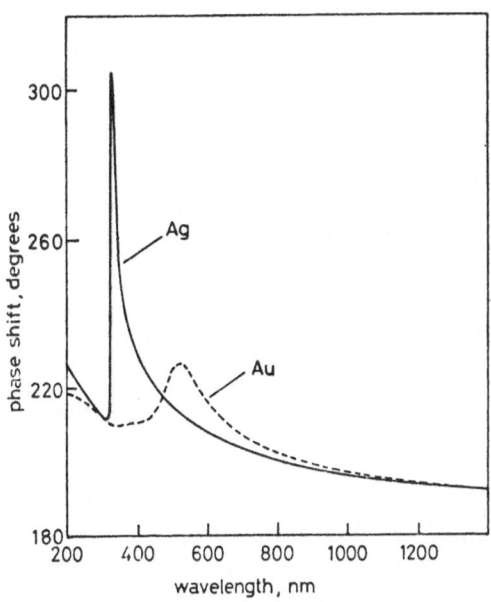

Fig. 10. The phase shift in a reflected electromagnetic wave on reflection *in vacuo* at normal incidence at silver and gold plane surfaces, calculated from refractive index data in [40].

the metal surface infrared selection rule [34], but at optical frequencies, particularly in the region of surface electronic resonances the phase shift in the parallel electric field component of an electromagnetic wave upon reflection at the surface is > 180°. This is shown in Fig. 10 for normal incidence on plane silver and gold surfaces. As a result of this phase lag the total parallel optical electric field component at the surface is not zero, or equivalently the polarization of the image of the molecule by the incident electromagnetic field lags that of the molecule itself by >180° at optical frequencies. The parallel electric field component may be relatively large at certain points on rough or particulate metal surfaces illuminated at the frequencies of surface conduction electron resonances. It is therefore our conclusion that the selection rule for SER scattering, which occurs at the frequencies of surface resonances (see below) cannot be obtained by the image field method of [39]. It seems probable indeed that the selection rules for SERS and for normal bulk-phase Raman scattering are the same, but it is necessary that the mechanism of the coupling of the molecular vibrations to the metal surface excitations be known before the SERS selection rule can be confidently proposed.

9.6 Some Observations of the Mechanism of the Surface Raman Enhancement

The investigation of the mechanism of SERS has received particular attention and a critical review of the several theories which have been proposed has been published by FURTAK and REYES [41]. Only a brief survey of some relevant experimental observations will therefore be given here. The main experimental features which it is necessary to encompass in the theory are the large Raman intensity enhancement of at least 10⁴-fold; the variation of SERS intensity with excitation wavelength; the crucial importance of surface roughness; the specificity of the enhancement to the surface monolayer; the present limitation to copper, silver and gold; the directionality of SER scattering from roughened bulk metal surfaces [12,13].

As already seen the SER scattering cross sections of typical adsorbate molecules are enhanced by 10^4-10^6 relative to the cross sections for normal Raman scattering, the exact enhancement deduced from the measurements depending on the value taken for the surface roughness factor. It has been observed by JEANMAIRE and VANDUYNE however that in the case of the dye molecule crystal violet adsorbed on silver there is a much larger total enhancement of *ca.* 10^9-fold relative to typical normal Raman scattering cross sections [9]. Crystal violet has a particularly large polarizability due to its electronic absorption in the visible range, as a result of which even the free molecule is a resonant Raman scatterer for excitation at normal visible wavelengths. The observation of the especially large Raman intensity for adsorbed crystal violet therefore strongly suggests that it is the polarizability of the adsorbate which is the significant molecular property for SER scattering. This conclusion is of course consistent with the result described in the previous section, that all of the vibrational symmetry classes of pyridine which are allowed by the polarizability selection rule for normal Raman scattering are also SERS-active.

It has been found that the intensity of SER scattering is very dependent on the excitation wavelength [42,43], and we have recently measured this dependence over a wide range of wavelengths for a number of adsorbates on roughened silver surfaces [23]. As shown in Fig. 11 there is a resonant scattering maximum for each adsorbate in the range 600-750 nm. These maxima are remarkably close together in wavelength in view of the varied nature of the molecules concerned, since as free molecules pyridine, triphenylphosphine, and cyanide ions are potential Raman scatterers only in the ultraviolet beyond 250 nm, whereas ruthenium red in solution has a pronounced resonance

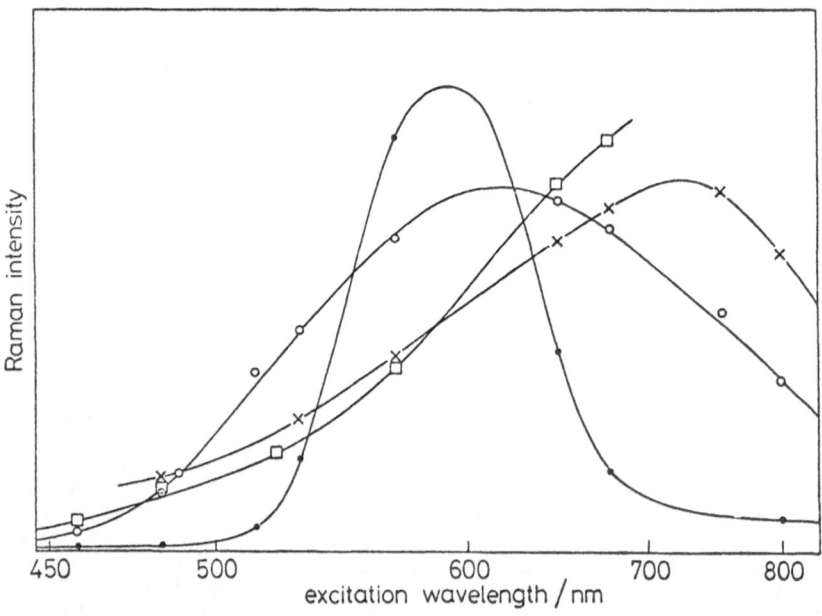

Fig.11. Raman excitation profiles for pyridine (✗ , 1008 cm^{-1}), triphenylphosphine (○ , 1000 cm^{-1}), cyanide (☐ , 2114 cm^{-1}), and ruthenium red ([Ru$_3$O$_2$(NH$_3$)$_{14}$]$^{6+}$, ● , 275 cm^{-1}) adsorbed at aqueous silver interfaces [23].

Raman scattering maximum for excitation at 530 nm. Preliminary results for adsorbates on copper and gold surfaces [23] suggest that there is a bigger difference in the wavelength of the SERS resonance maximum however for different metals. These observations add support to the proposals of MOSKOVITS [44], CREIGHTON *et al.* [29], BURSTEIN *et al.* [45], and OTTO and coworkers [20,46], that it is a resonant response of the electrons of the metal surface which is principally responsible for the greatly enhanced scattering. MOSKOVITS [44] and CREIGHTON *et al.* [29] consider the resonant response to be the excitation of surface plasma oscillations in the rough or particulate surfaces, which behave as resonant bodies or cavities at optical frequencies. At resonance the molecules at the surface experience an enhanced electromagnetic field and their Raman scattering by the normal induced dipole mechanism is therefore also enhanced. BURSTEIN *et al.* [45] and OTTO and coworkers [20, 46] are instead of the view that it is the excitation of electron-hole pairs which constitutes the response of the metal surface, and BURSTEIN *et al.* propose several non-classical coupling mechanisms between these metal surface excitations and the vibrations of the adsorbate molecules. The excitation of electron-hole pairs at extended metal surfaces, like that of surface plasmons, requires surface roughness so that momentum parallel to the surface can be conserved in the excitation, and both of these theories therefore account for the crucial importance for SERS of the surface roughness.

It has been pointed out by OTTO and coworkers [20] that the electromagnetic fields associated with surface plasmons extend out from the surface for many molecular diameters. The coupling of these fields to adsorbate motions is therefore weak, and moreover would not provide the specificity observed in SERS towards molecules only very close to the surface (Fig. 4). The mechanisms proposed [45] for coupling between molecular states and the more localized electron-hole surface excitations may on the other hand provide the required specificity. However measurements of the SER scattering by molecules adsorbed on metal diffraction gratings or on metal colloidal particles show that SER scattering has a dependence on geometrical variables which is similar to that for surface plasmon excitation at these surfaces, and thus appears to establish a link between SERS and surface plasmon-excitation. Thus metal diffraction gratings show sharp reflectance minima (Woods anomalies) for illumination at certain angles of incidence by light polarized perpendicular to the grooves. This absorption is due to the excitation of surface plasmons, which occurs only at a particular angle of incidence for a given periodic surface and incident wavelength, and the reflectance minimum is accompanied by a maximum in the re-emission of light in other directions due to surface plasmon decay [47]. PHILPOTT and coworkers have now shown that there is also a sharp maximum in the Raman scattering by molecules at the surface of a silver grating when it is illuminated at this angle with this incident polarization [48]. In an essentially similar experiment PETTINGER *et al.* used the technique of prism rather than grating coupling of the incident light to surface plasmons in a thin silver film, and also observed a sharp maximum in the SERS intensity from pyridine adsorbed on the silver film at the angle of incidence and with the correct polarization for which surface plasmon excitation was a maximum [49]. Colloidal silver or gold particles also show absorption of visible light due to the excitation of plasma oscillations. For spherical particles these oscillations take the form of spherical harmonics, and if the particles are very small compared to the wavelength it is only the dipolar oscillation which is excited, giving a distinct absorption band which results in the characteristic colours of silver and gold sols. As with diffraction gratings this absorption is accompanied by a scattering maximum at that wavelength due to plasmon excitation and decay, and increase in the particle size to dimensions comparable to the wavelength causes the absorption band to broaden and shift progressively to longer wavelengths and the scattering efficiency to increase. The SER scattering by pyridine adsorbed on colloidal silver and gold particles of various sizes has been investigated by CREIGHTON, BLATCHFORD and ALBRECHT [29], and it was shown that the SERS intensity is a maximum for excitation at the wavelength of the plasmon absorption, this SERS excitation maximum shifting smoothly with the plasmon absorption band to longer wavelengths with increase in particle size. This result may be seen in Fig. 12, which shows (full lines) the transmission spectra of silver sols of various particle sizes and (dashed lines) the SERS excitation profiles. The various particle sizes were obtained by aggregating very small particles, and it is the longer wavelength extinction maximum in each of the transmission spectra in Fig. 12 which is due to the aggregates, the extinction maximum near 400 nm in each spectrum being due at least in part to residual small particles.

Perhaps the most valuable aspect of these colloid observations is that they may provide a guide to the wavelength region in which a particular metal may show SER scattering [29]. Thus it has been found that with copper and gold surfaces the excitation wavelength must be longer than 570 nm [13] whereas with silver it need only be > 450 nm, while the lack of success with other metals using visible light may be due to the use of unfavourable excitation wavelengths for these metals. For a particular metal the dipole resonance of small isolated spherical particles occurs at a wavelength at which the real part of the metal dielectric function is equal to $-2n^2$, where n is the refractive index of the ambient medium. Reference to refractive index data [40] shows that copper, silver and gold are unusual in having small particle resonances in the visible range. For small isolated silver particles and copper and gold particles respectively these resonances are near 400 nm and 525 nm, but if it is assumed that the roughened bulk metal surfaces behave as though they were composed of many particles packed close together on the surface, a substantial shift of these particle resonances to longer wavelengths would be expected to occur due to dipolar coupling of adjacent particle resonators [44], thus accounting at least qualitatively for the region of the effective excitation wavelengths for these metals. Published refractive index data show that for most metals the dipole resonances of small isolated particles are in the far ultraviolet. These may be shifted into the visible range by dipole coupling effects and/or by growth of appropriately large particles. Our unpublished calculations show however that though large shifts in the plasmon resonance wavelength may be obtained by particle growth the resonance becomes extremely broad and shallow for large particles. It may therefore be that success with SERS scattering by such metals will only come from ultraviolet excitation. On the other hand it has been suggested [7] that the intensity of SER scattering parallels the reflectivity of the metal surface, and it follows that the most favourable region for SERS excitation with many metals may be in the infrared, where metal reflectivity is high. The experimental test of which of these wavelength regions is the more favourable outside the copper, silver, gold group remains to be done.

Other features of SERS, such as the strong directionality of SER scattering from roughened metal surfaces [12,13] and the differing effects of change of electrode potential on the intensity of different bands of the same adsorbate [9] also remain to be understood. Activity has accelerated greatly in this field during 1979-80 however, bringing in research groups with a wide variety of expertise, and rapid progress in unfolding the theory and applications of this promising new surface technique may be expected.

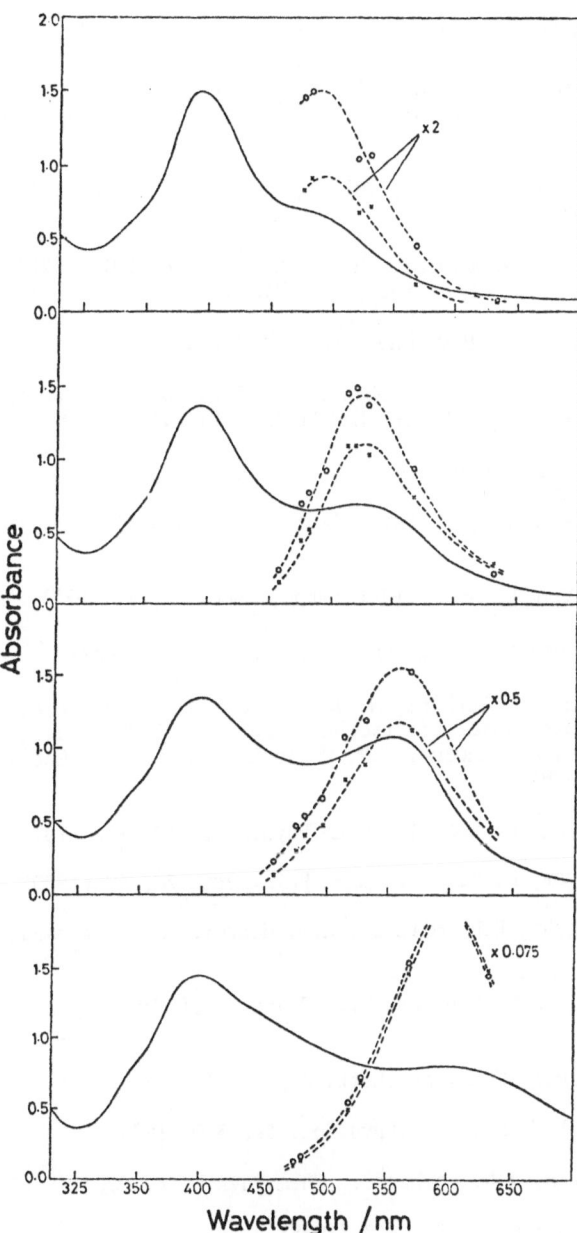

Fig. 12. The transmission spectra (full lines) of aqueous dispersions of silver particles of various sizes, and (dashed lines) the excitation profiles of the 1010 and 1038 cm^{-1} Raman bands of pyridine adsorbed on these colloidal particles [29].

1. J.R. Nestor, E.R. Lippincott: J.Raman Spectrosc. *1*, 305 (1973)

2. Y. Nishimura, A.Y. Hirakawa, M. Tsuboi; In *Advances in Infrared and Raman Spectroscopy*, Vol. 5, ed. by R.J.H. Clark, R.E. Hester (Heyden, London 1978) p.217

3. J.Behringer: In *Raman Spectroscopy*, Vol. 1, ed. by H.A. Szymanski (Plenum, New York 1967) p.168

4. H.Yamada, T. Amamiya, H. Tsubomura: Chem.Phys.Lett. *56*, 591 (1978); M. Fujihira, T. Osa: J.Amer.Chem.Soc. *98*, 7850 (1976)

5. T. Takenaka, T. Nakanaga: J.Phys.Chem. *80*, 475 (1976)

6. R.P. Van Duyne: In *Chemical and Biochemical Applications of Lasers*, Vol. 4, ed. by C.B. Moore (Academic Press, New York 1978) p.101

7. Physics Today, April 1980, p.18

8. M. Fleischmann, P.J. Hendra, A.J. McQuillan: Chem.Phys.Lett. *26*, 163 (1974)

9. D.L. Jeanmaire, R.P. Van Duyne: J.Electroanal.Chem. *84*, 1 (1977)

10. M.G. Albrecht, J.A. Creighton: J.Amer.Chem.Soc. *99*, 5215 (1977)

11. G. Blondeau, M. Froment, J. Zerbino, N. Jaffrezic-Renault, G. Revel: J.Electroanal.Chem. Interfacial Electrochem. *105*, 409 (1979); J.G. Bergman, J.P. Heritage, A. Pinczuc, J.M. Worlock, J.H. McFee: Chem. Phys.Lett. *68*, 412 (1979)

12. B. Pettinger, U. Wenning, H. Wetzel: Chem.Phys.Lett. *67*, 192 (1979)

13. U. Wenning, B. Pettinger, H. Wetzel: Chem.Phys.Lett. *70*, 49 (1980)

14. J.F. Evans, M.G. Albrecht, D.M. Ullevig, R.M. Hexter: J.Electroanal. Chem. *106*, 209 (1980)

15. R. Dornhaus, M.B. Long, R.E. Brenner, R.K. Chang, Surf.Sci. *93*, 240 (1980)

16. B. Pettinger, U. Wenning: Chem.Phys.Lett. *56*, 253 (1978)

17. D.A. Weitz, A.Z. Genack: Bull.Amer.Phys.Soc. *24*, 340 (1979)

18. G.R. Erdheim, R.L. Birke, J.R. Lombardi: Chem.Phys.Lett. *69*, 495 (1980)

19. T.E. Furtak: Solid State Comm. *28*, 903 (1978)

20. J. Billmann, G. Kovacs, A. Otto: Surf.Sci. *92*, 153 (1980)

21. R.P. Cooney, E.S. Reid, M. Fleischmann, P.J. Hendra: J.Chem.Soc. Faraday I, *73*, 1691 (1977); H.S. Gold, R.P. Buck: J. Raman Spectrosc. *8*, 323 (1979)

22. V.V. Marinyuk, R.M. Lazorenko-Manevich, Ya. M. Kolotyrkin: Dokl.Akad. Nauk SSSR *242*, 1382 (1978)

23. C.G. Blatchford, J.R. Campbell, J.A.Creighton: to be published

24. G. Hagen, B. Simic Glavaski, E. Yeager: J. Electroanal.Chem. *88*, 269 (1978)

25. T.H. Wood, M.V. Klein: J.Vac.Sci.Technol. *16*, 459 (1979)

26. R.R. Smardzewski, R.J. Colton, J.S. Murday: Chem.Phys.Lett. *68*, 53 (1979)

27. R.L.Paul, A.J. McQuillan, P.J. Hendra, M. Fleischmann, J.Electroanal. Chem. *66*, 248 (1975)

28. B.J. Bandy, D.R. Lloyd, N.V. Richardson: Surf.Sci. *89*, 344 (1979)

29. J.A. Creighton, C.G. Blatchford, M.G. Albrecht: J.Chem.Soc. Faraday II, *75*, 790 (1979)

30. R.P. Cooney, E.S. Reid, P.J. Hendra, M. Fleischmann: J.Amer.Chem.Soc. *99*, 2002 (1977); R.P. Cooney, M. Fleischmann, P.J. Hendra, J.Chem.Soc. Chem.Comm. 235 (1977)

31. R.P. Cooney, P.J. Hendra, M. Fleischmann: J.Raman Spectrosc. *6*, 264 (1977)

32. J.M. Stencel, E.B. Bradley: J.Raman Spectrosc. *8*, 203 (1979)

33. W. Krasser, A. Ranade, E. Koglin: J. Raman Spectrosc. *6*, 209 (1977); W. Krasser, A.J. Renouprez: J.Raman Spectrosc. *8*, 92 (1979)

34. H.A. Pearce, N. Sheppard: Surf.Sci. *59*, 205 (1976)

35. W. Ho, R.F. Willis, E.W. Plummer: Phys.Rev.Lett. *40*, 1463 (1978)

36. J.H.S. Green, W. Kynaston, H.M. Paisley: Spectrochim.Acta *19*, 549 (1963)

37. G.R. Erdheim, R.L. Birke, J.R. Lombardi: Chem.Phys.Lett. *69*, 495 (1980)

38. C.S. Allen, R.P. Van Duyne: Chem..Phys.Lett. *63*, 455 (1979).

39. R.M. Hexter, M.G. Albrecht: Spectrochim.Acta *35A*, 233 (1979)

40. H.J. Hagemann, W. Gudat, C. Kunz: J.Opt.Soc.Amer. *65*, 742 (1975)

41. T.E. Furtak, J.Reyes: Surf.Sci. *93*, 351 (1980)

42. J.A. Creighton, M.G. Albrecht, R.E. Hester, J.A.D. Matthew: Chem.Phys. Lett. *55*, 55 (1978)

43. B. Pettinger, U. Wenning, D.M. Kolb: Ber.Bunsenges.Phys.Chem. *82*, 1326 (1978)

44. M. Moskovits: J.Chem.Phys. *69*, 4159 (1978); Solid State Comm. *32*, 59 (1979)

45. E. Burstein, Y.J. Chen, C.Y. Chen, S. Lundquist, E. Tosatti: Solid State Comm. *29*, 567 (1979); C.Y. Chen, E. Burstein, S. Lundquist: Solid State Comm. *32*, 63 (1979)

46. A. Otto, J. Timper, J. Billmann, G. Kovacs, I. Pockrand: Surf.Sci. *92*, L55 (1980)

47. M.C. Hutley, D. Maystre: Optics Comm. *19*, 431 (1976); M.C. Hutley, V.M. Bird: Optica Acta *20*, 771 (1973)

48. A. Girlando, M.R. Philpott, D. Heitmann, J.D. Swalen, R. Santo: to be published

49. B. Pettinger, A. Tadjeddine, D.M. Kolb: Chem.Phys.Lett. *66*, 544 (1979)

10. Vibrations of Monatomic and Diatomic Ligands in Metal Clusters and Complexes – Analogies with Vibrations of Adsorbed Species on Metals

N. Sheppard

10.1 Background

Chemistry is concerned with the transformation of matter at the molecular level and catalysis by surfaces (heterogeneous catalysis) is of major importance in promoting such chemical reactivity. Metal surfaces are important catalysts for a wide variety of chemical transformations including, for example, reduction, oxidation, bond breaking and bond-forming processes [1]. A knowledge of the structures of the new chemical species formed by the interaction of reagents such as hydrogen, CO, N_2 or hydrocarbons with metal surfaces is therefore of great interest and potential importance.

During the past 5 to 10 years vibrational spectroscopy (in the forms of infrared, Raman or more recently, as high resolution electron energy loss spectroscopy, EELS) has re-emerged as a particularly important method for determining the structures of adsorbed species. Using one or another of these spectroscopies it is possible to determine the vibration frequencies, and from these often to deduce the molecular structures, of species adsorbed on finely divided metal catalysts [2,3]. The latter are usually supported on the surfaces of oxides such as silica and alumina; amongst other considerations, the oxide support helps to preserve the high areas of the metal catalysts at working reaction temperatures. More recent developments have enabled the study of species adsorbed as monolayers on single-crystal metal surfaces of known and specific atomic geometry. Such single-crystal work is increasingly being carried out in order to provide models to aid in the interpretation of the more complex results obtained on finely divided metal catalysts. The latter undoubtedly exhibit a variety of different surface sites or planes.

Using infrared and Raman spectroscopy it is equally possible to measure the vibrational spectra from ligands of analogous structures attached to metal clusters of known geometry. The latter can be prepared and purified in inorganic chemistry laboratories, and are often structurally analysed by x-ray crystallography. The spectra of such compounds provide guides for interpreting the spectra from adsorbed species on both finely divided and single-crystal metals.

A very considerable chemical literature already exists on the vibrational spectra of ligands attached to single metal atoms [4,5]. This is being rapidly expanded to include 'bridging' ligands attached to two, three or more metal atoms such as are likely to occur on metal surfaces.

It is the purpose of this brief paper to point to some of these literature sources and to give examples where the spectra of cluster compounds have successfully helped to determine or confirm the structures of chemisorbed species. The subject is a large one, particularly when hydrocarbon or other complex ligands are concerned. In this review we shall consider possible species from H_2, CO, N_2, O_2 and NO.

There has been discussion of the extent to which modes of vibration of ligands in metal transition complexes and cluster compounds will give similar wavenumber ranges to those for the analogous species adsorbed on the surfaces of metal particles or bulk metals. Some examples of such comparisons will be discussed below. At this point we should wish to make the general proviso that, because in a transition metal complex the ligand concerned may have its electronic structure, and hence its vibrations modified by electronic effects of other powerful ligands attached to the same or adjacent metal atoms, the range of wavenumbers appropriate to the vibrations of a ligand on a metal surface is likely to be considerably less than that exhibited by complexes or cluster compounds as a whole.

10.2 Hydrogen as Adsorbed Species or as a Ligand on Metals

It has long been known [1] that the reaction of hydrogen with metal surfaces normally results in the breaking of the H-H bond of the diatomic molecule. Either terminal M-H groups (M = metal) or bridging species, where the hydrogen interacts with more than one metal atom, or even interstitial hydrogen species are possible. Bridging species of the types HM_2 or HM_3, where the hydrogen interacts with two or three metal atoms respectively, are described as μ_2-H and μ_3-H ligands in the modern nomenclature of inorganic chemistry.

Much spectroscopic work has been done on the bond-stretching vibration frequencies, νMH, of individual MH groups. In a review in 1972 KAESZ and SAILLANT [6] gave characteristic wavenumber ranges for νM-H vibrations for different metals. The ranges quoted are wider than those likely to be applicable to hydrogen atoms bonded to metal surfaces because of the electronic effects of other powerful ligands such as CO, NR_3 or PR_3 (R = alkyl or aryl groups) attached to the same or adjacent metal atoms.

For νMH vibrations as a whole a range of 2300 cm^{-1} to 1700 cm^{-1} was shown to be appropriate [4,6], with the lower wavenumber end of the range being applicable when powerful electron-donating ligands are present or the complex has an overall negative charge. Even this rather general information was sufficient for the author to question with confidence an early assignment that had been made for two bands observed at 1251 and 1049 cm^{-1} by electron

energy loss spectroscopy (EELS) for hydrogen adsorbed on a
W(100) crystal face [7] and similar assignments for H on W(100),
W(111) and W(110) planes [8]. (This in no way diminished the
author's admiration for the technical achievements that had
enabled the bands for single monolayers of hydrogen to be measured at all!) The two bands for H on W(100) had been assigned
respectively to 'on-top' or terminal M-H bonds and bridged
M–H–M species respectively. The band of high wavenumbers at
1251 cm^{-1}, which was obtained at low surface coverage, seemed
much too low in wavenumber to be consistent with the presence
of non-bridged M-H groups, judged by the literature on coordination and cluster compounds [6].

At the time of KAESZ and SAILLANT'S review only fragmentary
data were available in the literature for the wavenumbers of
vibrations of bridging hydride ligands. However the suggested
range was 1400 to 800 cm^{-1}, and hence for the case of H on W(100)
our conclusion was that both bands had to be assigned to bridged
species. The lower wavenumber band at 1049 cm^{-1} had been shown
by the original authors [7] to be unambiguously assigned to a
bridged species as such adsorption sites were the only possible
ones for the known high surface coverage of two H atoms per unit
mesh.

The paucity of experimental information on bridged H species
led us, in this laboratory, to make a more systematic study of
the wavenumbers associated with vibrations of hydrogen species
of the type M–H–M [9]. Such a grouping has two modes of vibration associated with the stretching of MH bonds, and one 'angle-
bending' mode in which (in the presence of other ligands
or metal atoms) the H atom vibrates perpendicular to the plane of
the M_2H triangle. The forms of these vibrations are illustrated
below

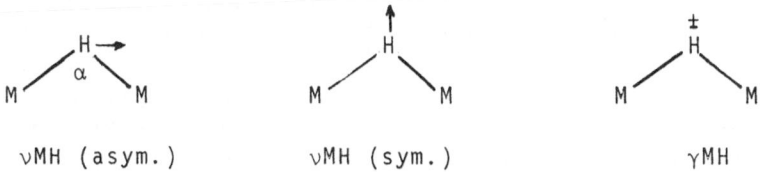

νMH (asym.)   νMH (sym.)   γMH

Although we were often able to successfully identify, with the
help of H to D substitution, two bands which could be associated
with νMH stretching modes, in some cluster compounds their wavenumbers were similar but in others well separated. It seemed
that these differences could be attributed to variable M–H–M
angles and we pointed out that, because the H atom is uniquely
light compared with the metal atoms, a simple central force-
field gives a theoretical ratio νMH(asym.)/νMH(sym.) = tan(α/2)
|9| where α denotes the MHM interbond angle. Comparison of our
experimental data with the angles ∝ determined in a few cases
through neutron diffraction, led to very good agreement with
the above equation. Qualitatively identical conclusions (based
on other experimental data) were reached independently by
KATOVIC and McCARLEY[10]. Given the above equation, the ratio

of the two wavenumbers can now be used to estimate the angle α
and - because the M-M distance in a cluster compound is often
known accurately from X-ray crystallographic studies - hence
the distance of the H atom above the M-M line.

The νMH(asym.) and νMH(sym.) wavenumbers coincide, according
to this model, for $\alpha = 90^{\circ}$. For $\alpha > 90^{\circ}$ the νMH(sym.) mode will
have the lower wavenumber, and vice-versa for $\alpha < 90^{\circ}$. In general
the νMH(sym.) wavenumber should move to lower wavenumber for
increasing α, given approximately similar M-H bonding. Because
of the operation of the 'metal surface selection rule'[11] , it
is the νMH(sym.) mode, which has a dipole charge perpendicular
to the metal surface, that is expected to be strongest at specular angles of reflection as was used in the W(100) measurements.
We therefore suggested [12] that the band at 1251 cm^{-1} observed
at lower coverage for H on W(100) came from a second bridged
species with a more acute value of α, and that this might be
associated with a metal surface reconstruction whereby some pairs
of W atoms come into closer contact. This was in analogy with
some similar findings for a bare W(100) surface at higher temperatures [13]. This explanation of the H/W(100) spectra seems
now to be generally accepted and additional non-specular EELS
studies by WILLIS [14,15] and IBACH [16] and their colleagues
have provided much information about the wavenumbers of the
νMH(asym.) modes, which have dipole charges parallel to the
surface. Hence deductions have been made about M$\frown$H$\frown$M angles
and in some cases the distances between the H atoms and the
metal surface plane. With the other surface planes W(111) and
W(110), it seems that EELS absorption between 1300 and 1250 cm^{-1}
[8] are consistently associated with μ_2-bridged hydrogens between
pairs of W atoms that are essentially in contact [12].

In a few cases δMH modes of terminal MH bonds have given
infrared bands between 800 and 600 cm^{-1} [5] and a few γMH modes
for μ_2-H species measured in this laboratory fall in the same
wavenumber range.

In this laboratory we have recently completed determinations
of the wavenumbers of the bands associated with the MH bond
stretching vibrations of a number of μ_3-H species in compounds
with H atoms bridging an equilateral triangle of metal atoms [17].
These have been interpreted in terms of the angle between the
M-H direction and the normal from the H atom to the M_3 plane,
using a theoretical treatment analogous to that described above
for the μ_2-H system. BARO, IBACH and BRUCHMANN [16] have measured, by EELS at specular and non-specular angles, both the
νMH(sym.) and the (doubly-degenerate) νMH(asym.) wavenumbers
for the vibrations of H on Pt(111). The ratio of these values,
with the help of the above theory, gives the distance of the adso
adsorbed H atom above the plane of the equilateral triangle of
Pt atoms to which it is attached [16, 17] .

In this laboratory we have also recently determined the
wavenumber (825 cm^{-1}) for a H atom in the centre of an octa-
hedron of metal atoms in [HRu$_6$(CO)$_{18}$]$^-$, [18]. This may provide
a model for detecting H atoms which have been absorbed into the

interstices between the first and second layers of metal atoms on a (111) face.

10.3 Carbon Monoxide as Adsorbed Species and Ligands

Carbon monoxide is a favourite ligand in transition-metal inorganic chemistry and literature on such vibrational spectra is vast [4,5]. However, a recent book by BRATERMAN [19] has provided a general but not comprehensive review of the vibrational spectra of metal carbonyls in small molecules. Also another more specific review by SHEPPARD and NGUYEN [20] has described the applicability of such spectral data to the identification of the species obtained from CO adsorbed on both finely divided and single-crystal metal surfaces. Using a modification of an assignment scheme originally proposed by EISCHENS and his colleagues [21,22], the following approximate wavenumber ranges have been proposed for CO molecules on metal surfaces bonded to one metal atom (the linear or terminal species) or to 2, 3 or 4 metal atoms (μ_2-CO, μ_3-CO and μ_4-CO species respectively).

Linear M-CO 2130 to ca. 2000 cm^{-1}

μ_2-bridged M_2(CO) ca. 2000 to 1880 cm^{-1}

μ_3-bridged M_3(CO)

and $\Big\}$ 1880 to 1650 cm^{-1}

μ_4-bridged M_4(CO)

It should be emphasised that the experimental data relating to the μ_3 and μ_4-species are fragmentary, and this range may be modified or sub-divided in the future. However, the wavenumber ranges for the linear and μ_2-species are well established. In a few recent studies by EELS of CO adsorbed on high-index planes (these exhibit regularly spaced steps between terraces of a low-index plane) it has been found that even lower wavenumbers for adsorbed CO can be found down to 1530 cm^{-1} [23].

The extensive experimental data for adsorbed CO enables a systematic comparison to be made of the wavenumber ranges associated with metal clusters and surface species. For linear MCO species, in general electron back-donation from the metal into the π anti-bonding orbital of a ligated CO molecule leads to a lowering of νCO. The extreme bonding possibilities can be represented in valence-bonding terms as in (1) and (2) below

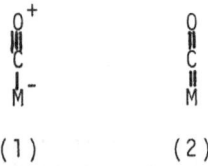

(1)   (2)

These effects are particularly marked in negatively or positively charged transition-metal complexes, but can also be substantial for other powerful ligands attached to the same metal atoms.

For adsorbed species, similar effects are seen in the wavenumber range associated with CO attached to metal ions (ca. 2200 to 2130 cm^{-1}) and to uncharged metals (ca 2130 to 2000 cm^{-1}). These compare with 2143 cm^{-1} for CO in the gas phase. However negatively charged metal atoms are not encountered on surfaces and there are no surface-based analogies to some of the lower νCO wavenumbers of linear species in complexes [4].

However, even when the abnormal effects of strongly electron-donating ligands are avoided, e.g. when considering metal cluster compounds with only CO ligands, systematic differences are found between the νCO wavenumbers and those from adsorbed species. For example, whereas the μ_2-bridged CO ligands in cluster compounds normally have νCO wavenumbers between 1900 and 1800 cm^{-1}, with the adsorbed species the typical wavenumbers are about 100 cm^{-1} higher [20]. This difference presumably reflects the net electron-donating properties of CO attached to metal compared with that of another metal atom (as in a metal particle) where the the net electron donation must be near zero (there could be some slight differential electronic effect between the outer or surface metal atoms and the inner ones in the metal particles).

Although the strength of the νCO modes in the infrared has led to them being most widely used for structural characterisation, the νM-C and νM-CO modes also give prominent features, the former tending to be stronger in the Raman effect, and the latter in the infrared. In inorganic complexes, for linear or terminal M-CO groupings, the wavenumber ranges associated with these modes appear to be approximately 500 to 350 cm^{-1} and 650 to 450 cm^{-1} respectively. In EELS studies of adsorbed species it is the νM-C modes that are expected to be stronger because of the operation of the 'metal surface selection rule'. The corresponding bands so far observed fit well into the above wavenumber range for νM-C. [4,5,20]

10.4 Nitrogen as Adsorbed Species and as Ligands

The nitrogen molecule (dinitrogen according to modern nomenclature when it occurs as a ligand) is isoelectronic with carbon monoxide. It is therefore not surprising that it forms linear or 'end-on' ligands and adsorbed species analogous to those obtained with CO. Possible extreme 'valence-bond' structures are

(3)   (4)   (5)   (6)

No 'end-on' $\mu_2$bridged species such as (5) have so far been identified either as adsorbed species or in cluster compounds. The rather different bridged species of type (6) has been characterised in complexes but is less likely to be applicable to surfaces except possibly for those of very open structure. The non-bridging 'end-on' species, of which (3) and (4) represent bonding extremes, have been extensively studied in inorganic complexes and have also been observed as an adsorbed species on finely divided metals. Although they have not yet been observed on single-crystal planes they have been observed on a polycrystal-line Pt(111) ribbon [24].

The first identification of an 'end-on' dinitrogen speices was made by EISCHENS and JACKNOW [25] in 1964 on a finely divided nickel catalyst using infrared spectroscopy. Shortly afterwards ALLEN and SENOFF [26] discussed it as a ligand in an inorganic coordination complex. The 'end-on' structure was postulated because of the intensity of the infrared bands in the triple bond region [25,26] and the structure was subsequently proved by the resolution of two $^{14}N^{15}N$ bands from a matrix isolated complex [27].

Because of the potential importance of such perturbed nitro-gen molecules in relation to the fixation of nitrogen, there are several extensive reviews of the overall literature [28-30]. Although these are not solely concerned with the vibration frequencies of the ligands, they do provide extensive lists of these data for the linear species. For the first-row transition metals the approximate range of $\nu N \equiv N$ wavenumbers in inorganic complexes is 2170 to 1860 cm^{-1}, to be compared with the gas-phase value for nitrogen of 2331 cm^{-1}. Somewhat lower values are obtained for some second and third row complexes.

In metal/N_2 complexes obtained through matrix isolation, a somewhat higher range of ca. 2250 to 2070 cm^{-1} is observed [4, 27,30]. This range is not dissimilar to the range 2260 to 2185 cm^{-1} [31-33] so far found for adsorbed species on finely-divided metals.

At room temperature the surface adsorbed species of this type are not strongly held and the strengths of the absorption bands are dependent on the pressure of gas-phase nitrogen [24,31-33] VAN HARDEVELD and his colleagues [31,32] postulate that adsorp-tion occurs on specific B_5 sites but this has not yet been verified for single-crystal planes - such sites occur, for example, as the between-row sites on (110) planes of a face-centred cubic metal or at certain step sites on high-index planes. The stronger bonding of end-on nitrogen ligands to metal atoms in transition-metal complexes is probably related to the effects of other ligands and higher effective oxidation states of the metal atom.

The symmetrical end-on bridging species (6) is formulated as a triple-bonded N_2 species because the wavenumbers of the absorp-tion bands can be close to those of the non-bridging species. The range so far observed seems to be between 2100 and 1650 cm^{-1} [30,34].

It is to be expected that on many metal surfaces adsorption of nitrogen will occur with dissociation. Transition metal complexes provide data for a nitrogen atom bonded to a single

metal atom by a triple bond, M≡N. The overall range appears
to be 1200 to 950 cm^{-1}, with the majority of absorptions falling
between 1100 and 1000 cm^{-1} [4,5].

As yet only fragmentary experimental data relate to M-N-M
and M$_3$N bonding situations [4]. Even these, because they refer
to linear and (possibly) planar bonding arrangements of nitrogen
atoms between the two and three metal atoms, may not be directly
relevant to surface species.

Little work has yet been done on the νM-N and δM-N≡N modes
of the 'end-on' species but there is some limited evidence that
they both fall in the general range 550 to 450 cm^{-1} for inorganic
complexes.[29,34] In some more weakly bound complexes observed
by matrix isolation techniques [27] νM-N modes are postulated
to occur down to 300 cm^{-1}. The lower wavenumber range may prove
to be more applicable to species adsorbed on metals.

10.5 Oxygen as Adsorbed Species and as Ligands

Various forms of transition-metal complexes involving O$_2$ species
are known |4,5|. They can usually be related to the diatomic
species O$_2$, O$_2^-$ (the superoxide ion) and O$_2^=$ (the peroxide ion).
These have formal bond orders of 2, 1½ and 1 respectively as the
added electrons fill π-antibonding orbitals. In vibration spectra
these species have wavenumbers of 1555 (O$_2$), ca. 1143 (O$_2^-$) and
ca. 1066 cm^{-1} (O$_2^=$) [4].

Complexes related to O$_2^-$ normally have direct bonding of the
metal atom to only one oxygen and the MO$_2$ entity is non-linear
as depicted in the two valence-bond extremes (7) and (8) below.

    (7)   (8)

These superoxide complexes appear to absorb in the range 1150
to 1100 cm^{-1} although there are not yet many examples of them
[4].

By contrast, whether observed in inorganic complexes [4] or
as oxygen/metal complexes in matrix isolation experiments [35,36]
it seems that O$_2$ ligands of the peroxide type are symmetrical
as in the alternative valence-bond formulations (9) or (10).

O———O   O———O
                      M^{2+}
    M

    (9)   (10)

The symmetry of these complexes has been shown both by X-ray crystallography and through spectra from mixed ^{16}O - ^{18}O species which show only one band. In transition-metal complexes the νOO modes occur rather consistently between 900 and 800 cm^{-1} [4,5] and there is some degree of coupling with νMO modes of the same symmetry; the latter normally occur between 580 and 470 cm^{-1} [4]. In the matrix-isolated species νOO occur in the somewhat higher wavenumber range of 1100 to 900 cm^{-1} [35,36], and the νMO modes occur at somewhat lower wavenumbers to inorganic complexes. Both these differences probably relate to a reduced perturbation of the original oxygen ligand through bonding to the metal atom. Peroxide-type O_2 species do not yet seem to have been detected on metal surfaces.

Bridged MOOM inorganic complexes of structures (11) or (12) are known although only the latter is likely to occur on a metal surface. The νOO wavenumber range seems to be approximately the

(11)   (12)

same as for the transition-metal MO_2 complexes [4].

It is of course to be expected that in many cases chemisorption of O_2 onto a metal surface will be dissociative in type to give oxygen atoms bonded to one or more metal atoms. When oxygen is bonded to a single metal atom, linkages approximating to double bonds M=O are expected. In inorganic complexes the general range of 1040 to 890 cm^{-1} is found for $\nu M=O$ when there is only one such bond per molecule or ion [5]. $\nu M-O$ vibrations, where an oxygen atom is bonded to more than one metal atom, can be expected in the general range of about 800 to 500 cm^{-1} when the M-O-M group is bent [5]. EELS results for O on Ni(100) [3] give bands which fall in the lower range 450 to 300 cm^{-1}. Here the oxygen atom is thought to be bonded to four metal atoms, and low wavenumbers may generally occur when O is bonded to more than two metal atoms. Linear MOM groups, less likely on surfaces, can have considerably higher wavenumbers than 800 cm^{-1} [5].

10.6 Nitric Oxide, NO, as Adsorbed Species and as Ligands

A large number of inorganic coordination complexes containing NO ligands have been studied by vibrational spectroscopy [4,5]. The reference states are the NO^+ ion, (isoelectronic with CO) with a formal triple bond which absorbs near 2200 cm^{-1}, and NO itself, with a bond order of $2\frac{1}{2}$, which absorbs at 1876 cm^{-1} in the gas phase. An NO^- species with a double bond would be expected to absorb between 1700 and 1600 cm^{-1}.

Like CO, NO is thought to bond to metal atoms as a σ-donor and a π-acceptor. In view of the different possible oxidation states for the molecule it is not surprising to find νNO covering a rather

wide wavenumber range of from about 2000 to 1500 cm^{-1} [4,5] .
X-ray studies have shown the presence of both linear and bent
M-NO groups, the former approximating to a triple-bonded ligand
and the latter to a double-bonded one as in (13) and (14) below.

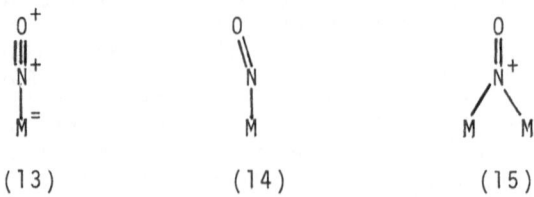

(13)   (14)   (15)

In some cases both types of group occur in the same molecule and
it is then found that there are two different NO frequencies,
one in the upper half of the above range and the other in the
lower half [4]. νNO adsorbed on metal catalysts usually absorbs
between 1900 and 1800 cm^{-1}, i.e. well within the range for complexes, although in one case an absorption as high as 2190 cm^{-1}
has been observed [38].

νMN and δMNO vibrations fall relatively close together in
the wavenumber range of about 650 to 450 cm^{-1}; these two group
motions are probably considerably mixed by coupling when the
MNO group is non-linear.

In a very few cases bridged NO ligands as in (15) have been
observed and these absorb between 1520 and 1480 cm^{-1} [39]; when
more examples are known the range will probably be extended.

Because NO is an odd-electron molecule there are possibilities of dimerization in a metal complex with a formula accommodat-
ing two NO groups, and presumably also on surfaces. If these
in turn bond to metal they are likely to give νNO wavenumbers
considerably below 1500 cm^{-1} [4,5].

10.7 General Conclusions

There has been considerable discussion as to whether or not metal
complexes and clusters provide good analogies for the species
observed or expected through adsorption on metal catalyst
surfaces [40-42] . We are particularly concerned here with the
vibrational spectroscopic manifestations of such ligands or
adsorbed species. In the last analysis, whether or not the
analogies are useful ones depends on judgements made by individual chemists in selecting data from complexes for comparison with
those from adsorbed molecules. Matrix-isolated species fall
somewhat between the two categories.

It seems to be rather clear, from the examples cited above,
that a given ligand within transition-metal complexes or
clusters usually show a considerably wider range of wavenumbers
for a given mode of vibration, than do the analogous species
adsorbed on surfaces. This difference is most probably caused
by the wider range of powerful co-ligands attached to the same

or adjacent metal atoms in the inorganic complexes. Usually the νXY modes (X,Y = C,N,O) for adsorbed species have average wavenumbers that are higher than the corresponding average for the complexes and closer to the value for the unperturbed molecule. There also seems to be a general expectation, on empirical as well as on theoretical grounds, that the νMX, νMY modes will fall at rather lower wavenumbers for the adsorbed species. However more evidence is needed before this can be accepted as a valid generalisation.

Matrix-isolated species tend to absorb' in positions not far from those observed with adsorbed species. To date the majority of those studied have been of the type ML_n (L = ligand, n = 1 to 4) but in a few cases [42-44] M_nL species have been observed. These are likely to give even closer analogies to adsorbed species, and also in such compounds vibrations of bridged species can be studied. However the advantage of matrix work is somewhat offset by the difficulties in identifying the spectra of particular species with the same or different values of n, within the overall mixture. If inorganic complexes differ more from adsorbed species in their electronic structures, there is at least the compensating advantage that individual compounds can be isolated and if necessary have their structures determined by X-ray crystallography.

In many of the formal valence-bond formulations of metal/ligand species given above the metal atom bears a positive or negative charge. However if the metal atom is part of a larger perhaps 3-dimensional, cluster there will be opportunities for the charge to be delocalised. The metal clusters might be thought of as 3-dimensional analogies to the aromatic rings of organic chemistry. They can not only delocalise charge densities but they can also transmit electronic effects from adjacent adsorbed species or ligands. In aromatic compounds a given substituent will usually have a recognisable range of wavenumbers for a given mode of vibration, but that relatively narrow wavenumber range may reflect a much wider range of rates of reaction or equilibrium constants involving that group. This is because small percentage changes in large energies of activation or enthalpies of reaction can have very large effects on rate or equilibrium constants. In the same way we may expect characteristic vibration frequencies to be a considerable help in relating structures of ligands and adsorbed species, even although the reactivities are markedly different. In a sense that is just what catalysis is about!

Finally we may note that vibrational spectroscopy only provides a straightforward and powerful link when the same species does in fact occur in complexes and on surfaces. The detailed architecture of metal-atom arrangements on metal surfaces can be expected to often give rise to species for which no prior analogies are to be found amongst complexes and clusters. This is particularly so when bonding to several metal atoms is involved. In such cases the vibrational spectroscopist has to use even more imagination in drawing structural conclusions.

The author thanks Michael Chesters, David Chenery and many spectroscopic friends for stimulating discussions which have contributed to some of the ideas expressed in this article. He is also much indebted to the Science Research Council, London, for a Senior Fellowship and for supporting his researches in this area over a considerable period of time.

References

1. G.C. Bond, 'Catalysis by Metals', Academic Press, New York, 1962.
2. L.H. Little, 'Infrared Spectra of Adsorbed Species', Academic Press, London, 1966.
3. J. Pritchard in 'Surface and Defect Properties of Solids', (M.W. Roberts and J.M. Thomas, eds.) Vol. I, Specialist Periodical Reports, Chemical Society, London, 1972, p.222 also in Vol. 7 (new termed 'Chemical Physics of Solids and their Surfaces') 1978, p.157.
4. K. Nakomoto, 'Infrared and Raman Spectra of Inorganic and Coordination Compounds', (3rd. Ed.) Wiley/Interscience, New York, 1978.
5. D.M. Adams, 'Metal-Ligand and Related Vibrations', Arnold, London, 1967.
6. H.D. Kaesz and R.B. Saillant, Chem. Rev., $\underline{72}$, 231 (1972).
7. H. Froitzheim, H. Ibach and S. Lehwald, Phys. Rev. Letters, $\underline{36}$, 1549 (1976).
8. C. Backx, B. Feuerbacher, B. Fitton and R.F. Willis, Phys. Letters, $\underline{60A}$, 145 (1977).
9. M.W. Howard, U.A. Jayasooriya, S.F.A. Kettle, D.B. Powell and N. Sheppard, J. Chem. Soc., Chem. Comm., $\underline{18}$ (1979).
10. V. Katovic and R.E. McCarley, Inorg. Chem., $\underline{17}$, 1268 (1978).
11. H. Ibach, Surface Science, $\underline{66}$, 56 (1977).
12. U.A. Jayasooriya, M.A. Chesters, M.W. Howard, S.F.A. Kettle, D.B. Powell and N. Sheppard, Surface Science, $\underline{93}$, 526 (1980).
13. M.K. Debe and D.A. King, Phys. Rev. Letters, $\underline{39}$, 708 (1977).
14. W. Ho, R.F. Willis and E.W. Plummer, Phys. Rev. Letters, $\underline{40}$, 1463 (1978).
15. M.R. Barnes and R.F. Willis, Phys. Rev. Letters, $\underline{41}$, 1729 (1978).
16. A.M. Baró, H. Ibach and H.D. Bruchmann, Surface Science, $\underline{88}$, 384 (1979).
17. J.A. Andrews, U.A. Jayasooriya, I.A. Oxton, D.B. Powell, N. Sheppard, P.F. Jackson, B.F.G. Johnson and J. Lewis, Inorg. Chem., (in press).
18. I.A. Oxton, S.F.A. Kettle, P.F. Jackson, B.F.G. Johnson and J. Lewis, J.C.S. Chem. Comm. 687 (1979).
19. P.S. Braterman, 'Metal Carbonyl Spectra', Acad. P., New York, 1975.
20. N. Sheppard and T.T. Nguyen in 'Advances in Infrared and Raman Spectroscopy', (R.J.H. Clark and R.E. Hester, eds.) Heyden, London, 1978, Vol. 5, p.67.
21. R.P. Eischens, W.A. Pliskin and S.A. Francis, J. Chem. Phys. $\underline{22}$, 1786 (1954).
22. R.P. Eischens and W. A. Pliskin, Adv. Catal., $\underline{10}$, 1 (1958).
23. Erley, H. Ibach, S. Lehwald and H. Wagner, Surface Science, 83, 585 (1979).

24. R. Shigeishi and D.A. King, Surface Science, 62, 379 (1977).
25. R.P. Eischens and J. Jacknow, Proc. 3rd. Intern. Congress on Catalysis, North-Holland Publishing Co., Amsterdam, 1964, p. 638.
26. A.D. Allen and C.V. Senoff, Chem. Comm., 621 (1965).
27. G.A. Ozin and A.V. Voet, Canad. J. Chem., 51, 3332 (1973).
28. A.D. Allen, R.O. Harris, B.R. Loescher, J.R. Stevens and R.N. Whiteley, Chem. Rev., 73, 11 (1973).
29. D. Sellman, Angew. Chem. (Eng. Ed) 13, 639 (1974).
30. J. Chatt, J.R. Dilworth and R.L. Richards, Chem. Rev., 78, 589 (1978).
31. R. van Hardeveld and A. van Montfoort, Surface Science, 4, 396 (1966).
32. R. van Hardeveld and F. Hartog, Adv. Cat., 22, 75 (1972).
33. A. Ravi, D.A. King and N. Sheppard, Trans. Faraday Soc., 64, 3358 (1968).
34. M.W. Bee, S.F.A. Kettle and D.B. Powell, Spectrochim. Acta, 30A, 585 (1974); 31A, 89 (1975).
35. E.P. Kündig, M. Moskovits and G.A. Ozin, Canad. J. Chem., 51, 2710 (1973).
36. D. McIntosh and G.A. Ozin, Inorg. Chem., 15, 2869 (1976).
37. S. Andersson, Solid State Comm., 24, 183 (1977).
38. S.V. Batychko, M.T. Rusov and L.M. Roev, Doklady Phys. Chem. (English ed.), 191, 328 (1970).
39. J.R. Norton, J.P. Collman, G. Dolcetti and W.T. Robinson, Inorg. Chem., 11, 382 (1972).
40. E.L. Muetterties, T.N. Rodin, E. Band, C.F. Brucker and W.R. Pretzer, Chem. Rev., 79, 91 (1979).
41. E.L. Muetterties, Bull. Soc. Chim. Belg., 84, 959 (1975) and 85, 451 (1976); Angew. Chem. (Eng.), 17, 545 (1978).
42. M. Moskovits, Acc. Chem. Res., 12, 229 (1979).
43. M. Moskovits and J.E. Hulse, Surface Science, 61, 302 (1976).
44. J.E. Hulse and M. Moskovits, Surface Science, 57, 125 (1976).

11. Coupling Induced Vibrational Frequency Shifts and Island Size Determination: CO on Pt {001} and Pt {111}

D. A. King

With 3 Figures

11.1 Introduction

In both the early transmission infrared spectra from carbon monoxide chemisorbed on supported platinum catalysts [1], and more recent reflection absorption infrared spectra from Pt{111} recrystallised ribbons [2-5] and single crystals, [6,7] significant shifts of the C-O stretching frequency have been reported as a function of coverage. For CO on Pt{111} at 300 K, these independent studies have now confirmed that a band first appears at $\sim$2065 cm^{-1} and shifts to $\sim$2100 cm^{-1} at saturation. These shifts were originally ascribed to one of two possible factors: BLYHOLDER [8] proposed that a reduction of dπ^* backbonding (into the antibonding 2π^* orbital of CO) occurred as the coverage was increased, due to competition for metal d-electrons; while HAMMAKER et al [9] proposed that the shift was due to dipole coupling between CO molecules aligned parallel to each other on the surface, supporting their model with results from ^{12}CO/^{13}CO isotopic mixtures. CROSSLEY and KING [3], using the theory developed by HAMMAKER et al[9], showed that a ^{12}CO molecule coupled very weakly into a ^{13}CO environment (due to the relatively large difference in the singleton, or isolated molecule, C-O stretching frequencies for the two isotopes) and hence performed an experiment which conclusively demonstrated that the entire frequency shift observed for CO on Pt{111} was attributable to coupling effects, and not to chemical bonding effects. Using variable ^{12}CO/^{13}CO isotopic mixtures, it was demonstrated that the entire frequency shift could be reproduced at a constant total coverage, corresponding to saturation at 300 K, as the ^{12}CO composition increased from 0 to 100%. This demonstration was independent of the details of the theory. More recently, the theory has been significantly refined,[10,11,12] and it would now appear that the observed frequency shift is entirely compatible with dipole coupling theory. In subsequent work, outlined below, we have attempted to show how high resolution vibrational spectroscopy can be used to determine island sizes in chemisorbed overlayers [4]. This depends on the use of isotopes to determine the true singleton frequency of the adsorbed species.

11.2 CO on Pt{001}: Island Growth [4]

The infrared absorption bands observed during CO chemisorption on Pt{001}at temperatures of 160, 300 and 340 K are shown in figure 1. Although coverage induced frequency shifts are observed at all temperatures, it is noteworthy that the extent of the shift increases with increasing substrate temperature; in particular, the frequency at which the band first appears is temperature dependent. At 160 K, the band is first observed at 2088 cm^{-1}; at 300 K, at 2080 cm^{-1}; and at 340 K the band is first observed at 2075 cm^{-1}. The situation on Pt{001} is complicated by the fact that as the CO coverage is increased the surface Pt atoms undergo a reconstructive transition from a (5 x 20) to a (1 x 1) structure. Furthermore, at high coverages bridged CO species, with absorption bands at 1887 cm^{-1}(at a fractional coverage θ = 6.7) shifting to 1952 cm^{-1} (at θ = 0.75), have been observed [13] in electron energy loss spectra, these species being coadsorbed with the species producing the high frequency band shown in figure 1. At temperatures below about 340 K, however,

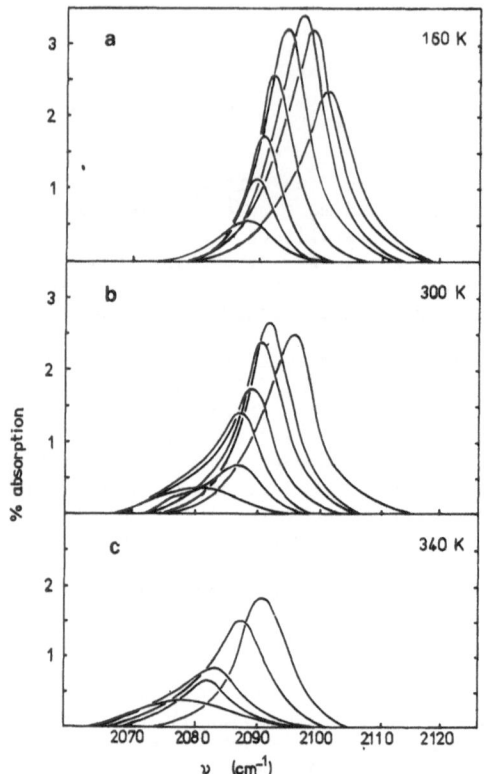

Fig. 1 Development of the IR band with increasing CO coverage on Pt{001}at three different substrate temperatures [4].

adsorption initially takes place exclusively into the high frequency, presumably linearly bonded, species, without conversion of the surface to the (1 x 1) structure. We can therefore concentrate here on the interpetation of the substrate temperature dependence of the absorption band at low coverages ($\lesssim 10^{14}$ molecules cm^{-2}) where these complications can be ignored.

The isotope mixture method was used [4] to determine the singleton frequency for ^{12}CO on Pt{001}, at a temperature of 300 K. The observed spectra, obtained at saturation coverage, are shown in figure 2, clearly demonstrating the characteristic features of a coupled overlayer system. At low ^{12}CO compositions two bands are clearly distinguished; in the conceptually simple two-body model, the high frequency band is attributed to the in-phase ^{12}CO - ^{12}CO and ^{12}CO - ^{13}CO coupled modes, while the low frequency band contains the in-phase ^{13}CO - ^{13}CO and the (very weak) out-of-phase ^{12}CO - ^{13}CO modes. The concurrent shifts in frequency of the low frequency band to lower frequencies and the high frequency band to higher frequencies as the ^{12}CO

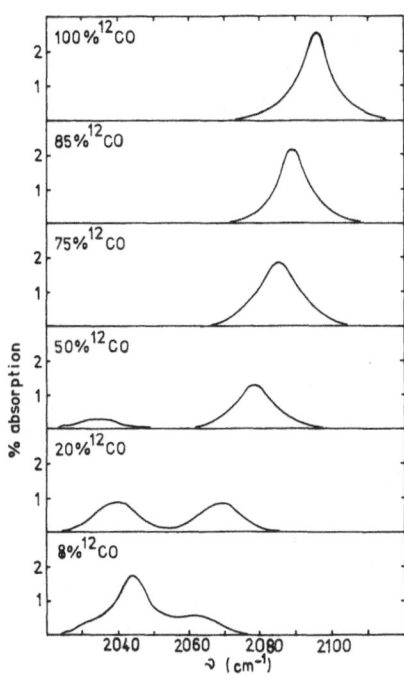

Fig.2 IR spectra from ^{12}CO - ^{13}CO mixtures on Pt{001} at 300 K and saturation coverages [4].

composition is increased, and the intensity stealing into the high frequency band (the 50-50 mixture shows much more intensity in the high frequency band) are clear indicators of a strongly coupled system. In particular, the singleton frequency ν_s for ^{12}CO is obtained, to within a few cm^{-1}, by extrapolating the shift in the high frequency band to 0% ^{12}CO, giving ν_s = 2062 cm^{-1}.

Isolated chemisorbed ^{12}CO species should therefore give an absorption band at ~2062 cm^{-1}. As shown in figure 1, however, at substrate temperatures of 160 to 340 K the band is always first observed at significantly higher frequencies. The obvious conclusion is that at low coverages the CO molecules segregate into islands, so that even at the lowest coverages the observed absorption band corresponds to the collective modes of chemisorbed CO coupled into the island environment. Assuming that the coupling does in fact arise from dipole interactions, which have a R^{-3} dependence, (where R is the distance between species) using the singleton frequency from the isotope experiments (hence assuming that the (5 x 20) to (1 x 1) conversion at high coverages is relatively unimportant), and also assuming a local structure within each island, the average number of molecules per island at a given coverage can be derived from the observed band frequency. The derived substrate temperature dependence of island size at a coverage of 10^{14} molecules cm^{-2} (θ = 0.075) for CO on Pt{001} is shown in Table 1. Of course, it would be desirable in future work to correlate both the local structure and the island size with a LEED study.

Island formation clearly implies the existence of attractive interactions between adsorbed species. However, the results in Table 1 contradict the expected equilibrium behaviour of a system with attractive interactions: a two-phase system is anticipated, with isolated species in equilibrium with

Table 1 Average island sizes, given as the number of molecules per island (N/n) as a function of substrate temperature T at a coverage of 10^{14} molecules cm^{-2}, derived from the C-O stretching frequency ν.

T[K]	$\nu (cm^{-1})$	N/n
160	2088	66
300	2082	24
340	2075	15
400	2072	11

congregated species. The observed behaviour implies the absence of a phase boundary. We have rationalised this using a simple model [4], in which the configurational entropy gained by subdividing N molecules cm^{-2} into n islands per cm^{-2} is counterbalanced by a decrease in energy as the number of molecules in the island boundaries (where nearest neighbours are missing) is increased with an increase in the number of islands. The model is consistent with the data in Table 1, yielding an attractive pairwise interaction energy between nearest neighbour chemisorbed species of 6 kJ mol^{-1}. This value is, in turn, consistent with van der Waals interactions between chemisorbed CO species.

11.3 CO on Pt{111}: "Gaseous" and Island Species

Several independent reflection-absorption infrared studies of CO on Pt{111} have now been reported [2-7], and there is quite good agreement between the sets of data. The major features are illustrated in figure 3, from the data of CROSSLEY and KING [4]. A band due to the C-O stretching frequency, from a linearly chemisorbed species, is first observed, at a substrate temperature of 300 K, at 2065 cm^{-1}, shifting to 2070 cm^{-1} at 0.3 x 10^{14} molecules cm^{-2}. Over the coverage range 0.4 to 1.6 x 10^{14} molecules cm^{-2}, however, the band has two overlapping components, the second appearing at 2083 cm^{-1} when the first is at 2074 cm^{-1}. The first band becomes less intense as the second takes over, and disappears at a coverage of 1.8 x 10^{14} molecules cm^{-2}, when the high frequency component is at 2090 cm^{-1}. This component shifts to higher frequencies with increasing coverage, reaching 2101 cm^{-1} at saturation. $^{12}CO/^{13}CO$ results at saturation coverage are entirely analogous to those for Pt{001} (figure 2), yielding a singleton frequency ν_s of 2065 cm^{-1}, very close to that for Pt{001}. The most important difference between the two systems however, is the coincidence between ν_s and the position of the ^{12}CO band first observed at 300 K (figure 3).

At coverages up to 0.4 x 10^{14} molecules cm^{-2}, CO adsorption on Pt{111} therefore occurs into a random or 'gaseous' state, with isolated species. At intermediate coverages, however, these free species are in equilibrium with islands, producing the second high frequency band; at higher coverages the free species are completely consumed at the expense of the islands. At lower temperatures, HORN and PRITCHARD [6] noted that the absorption band was first detected at frequencies significantly above ν_s for this system, suggesting the formation of relatively large islands at very low coverages, consistent with the CO/Pt{001} system.

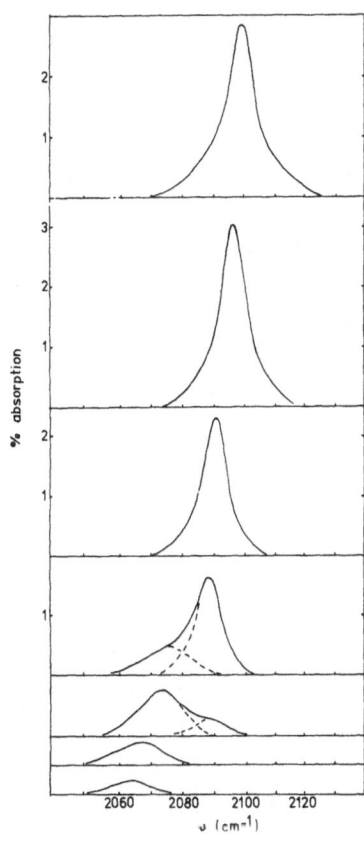

Fig.3 Development of the IR band with increasing CO coverage on Pt{111} at 300 K [4].

11.4 Conclusion

Dipole coupling between chemisorbed species in an overlayer produces coupled modes, and hence an observed IR absorption band which may be shifted considerably from that for isolated, or singleton, chemisorbed species. The singleton frequency can be obtained from mixtures of isotopes, and the observed band frequency from the single isotope experiment can then be used to infer details of the environment of chemisorbed species. Results from CO on Pt{001} demonstrate the feasibility of using this technique to determine average island sizes during chemisorption. The experiments demand a relatively high resolution ($\sim$10 cm^{-1}, or 1 meV) presently only obtainable with reflection absorption infrared spectroscopy.

References

1. R.P. Eischens and W.A. Pliskin, Advan. Catalysis 10 (1958) 1.

2. R.A. Shigeishi and D.A. King, Surface Sci. 58 (1976) 379.

3. A. Crossley and D.A. King, Surface Sci. 68 (1977) 528.

4. A. Crossley and D.A. King, Surface Sci. 95 (1980) 131.

5. M. Chesters, personal communication.

6. K. Horn and J. Pritchard, J. Physique $\underline{38}$ (1977) C4, 164.

7. H. Krebs and H. Lüth, Appl. Phys. $\underline{14}$ (1977) 337.

8. G. Blyholder, J. Phys. Chem. $\underline{68}$ (1964) 2772.

9. R.A. Hammaker, S.A. Francis, and R.P. Eischens, Spectrochimica Acta, $\underline{21}$ (1965) 1295.

10. G.D. Mahan and A.A. Lucas, J. Chem. Phys. $\underline{68}$ (1978) 1344.

11. M. Scheffler, Surface Sci. $\underline{81}$ (1979) 562.

12. B.N.J. Persson and R. Ryberg, to be published.

13. G. Pirug, H. Hopster and H. Ibach, to be published.

Dynamics of Solids and Liquids by Neutron Scattering

Editors: S. W. Lovesey, T. Springer
1977. 156 figures, 15 tables. XI, 379 pages
(Topics in Current Physics, Volume 3)
ISBN 3-540-08156-9

Contents:

Electron Spectroscopy for Surface Analysis

Editor: H. Ibach
1977. 123 figures, 5 tables. XI, 255 pages
(Topics in Current Physics, Volume 4)
ISBN 3-540-08078-3

Contents:

M. A. Van Hove, S. Y. Tong

Surface Crystallography by LEED

Theory, Computation and Structural Results

1979. 19 figures, 2 tables. IX, 286 pages
(Springer Series in Chemical Physics, Volume 2)
ISBN 3-540-09194-7

Contents:

Inelastic Electron Tunneling Spectroscopy

Proceedings of the International Conference, and Symposium on Electron Tunneling, University of Missouri-Columbia, USA, May, 25–27, 1977

Editor: T. Wolfram
1978. 126 figures, 7 tables. VIII, 242 pages
(Springer Series in Solid-State Sciences, Volume 4)
ISBN 3-540-08691-9

Contents:

Springer-Verlag
Berlin
Heidelberg
New York

Interactions on Metal Surfaces

Editor: R. Gomer
1975. 112 figures. XI, 310 pages
(Topics in Applied Physics, Volume 4)
ISBN 3-540-07094-X

Contents:
J. R. Smith: Theory of Electronic Properties of Surfaces. – *S. K. Lyo, R. Gomer:* Theory of Chemisorption. – *L. D. Schmidt:* Chemisorption: Aspects of the Experimental Situation. – *D. Menzel:* Desorption Phenomena. – *E. W. Plummer:* Photoemission and Field Emission Spectroscopy. – *E. Bauer:* Low Energy Electron Diffraction (LEED) and Auger Methods. – *M. Boudart:* Concepts in Heterogeneous Catalysis.

Monte Carlo Methods in Statistical Physics

Editor: K. Binder
1979. 91 figures, 10 tables. XV, 376 pages
(Topics in Current Physics, Volume 7)
ISBN 3-540-09018-5

Contents:
K. Binder: Introduction: Theory and "Technical" Aspects of Monte Carlo Simulations. – *D. Levesque, J. J. Weis, J. P. Hansen:* Simulation of Classic Fluids. – *D. P. Landau:* Phase Diagrams of Mixtures and Magnetic Systems. – *D. M. Ceperley, M. H. Kalos:* Quantum Many-Body Problems. – *H. Müller-Krumbhaar:* Simulation of Small Systems. – *K. Binder, M. H. Kalos:* Monte Carlo Studies of Relaxation Phenomena: Kinetics of Phase Changes and Critical Slowing Down. – *H. Müller-Krumbhaar:* Monte Carlo Simulation of Crystal Growth. – *K. Binder, D. Stauffer:* Monte Carlo Studies of Systems with Disorders. – *D. P. Landau:* Applications in Surface Physics.

Secondary Ion Mass Spectrometry SIMS-II

Proceedings of the Second International Conference on Secondary Ion Mass Spectrometry (SIMS II)
Stanford University, Stanford, California, USA, August 27–31, 1979
Editors: A. Benninghoven, C. A. Evans Jr., R. A. Powell, R. Shimizu, H. A. Storms
1979. 234 figures, 21 tables. XIII, 298 pages
(Springer Series in Chemical Physics, Volume 9)
ISBN 3-540-09843-7

Contents:
Fundamentals. – Quantitation. – Semicounductors. – Static SIMS. – Metallurgy. – Instrumentation. – Geology. – Panel Discussion. – Biology. – Combined Techniques. – Postdeadline Papers.

Theory of Chemisorption

Editor: J. R. Smith
1980. 116 figures, 8 tables. XI, 240 pages
(Topics in Current Physics, Volume 19)
ISBN 3-540-09891-7

Contents:
J. R. Smith: Introduction. – *S. C. Ying:* Density Functional Theory of Chemisorption on Simple Metals. – *J. A. Appelbaum, D. R. Hamann:* Chemisorption on Semiconductor Surfaces. – *F. J. Arlinghaus, J. G. Gay, J. R. Smith:* Chemisorption on d-Band Metals. – *B. Kunz:* Cluster Chemisorption. – *T. Wolfram, S. Ellialtioğlu:* Concepts of Surface States and Chemisorption on d-Band Perovskites. – *T. L. Einstein, J. A. Hertz, J. R. Schrieffer:* Theoretical Issues in Chemisorption.

Springer-Verlag
Berlin
Heidelberg
New York